Stereoelectronic Effects

A. J. Kirby

University Chemical Laboratory
Cambridge

OXFORD
UNIVERSITY PRESS

OXFORD
UNIVERSITY PRESS

Great Clarendon Street, Oxford OX2 6DP
Oxford University Press is a department of the University of Oxford.
It furthers the University's objective of excellence in research, scholarship,
and education by publishing worldwide in
Oxford New York
Auckland   Cape Town Dar es Salaam   Hong Kong   Karachi
Kuala Lumpur Madrid Melbourne Mexico City   Nairobi
New Delhi Shanghai Taipei Toronto
With offices in
Argentina Austria Brazil Chile Czech Republic France Greece
Guatemala Hungary Italy Japan South Korea Poland Portugal
Singapore Switzerland Thailand Turkey Ukraine Vietnam

Oxford is a registered trade mark of Oxford University Press
in the UK and in certain other countries

Published in the United States
by Oxford University Press Inc., New York

ISBN 978-0-19-855893-4

Series Editor's Foreword

The key to understanding organic chemistry is the ability to recognise the controlling factors, stereoelectronic effects, that determine the reactions and shapes of molecules. A grasp of the relatively simple principles of stereoelectronic effects makes the vast array of organic reactions predictable and hence should be an essential part of all modern organic chemistry courses.

In this Primer, Tony Kirby provides an excellent account of the all-pervasive nature of stereoelectronic effects, which will be useful to students from the very start of, as well as throughout, their University chemistry course. Oxford Chemistry Primers have been designed to provide concise introductions relevant to all students of chemistry and contain only the essential material that would usually be covered in an 8–10 lecture course. These constraints mean that pericyclic reactions, which are obviously under stereoelectronic control, are not covered in this Primer, however there will be another Primer later in the series that deals explicitly with this topic.

This Primer on Stereoelectronic Effects will be of interest to apprentice and master chemist alike.

Dr Stephen G. Davies
The Dyson Perrins Laboratory, University of Oxford

Preface

Learning organic chemistry is hard work. The amount of information 'out there' is as close to infinite as makes no difference to even the most serious student, so we need ways of making the material digestible. The best solution is to make sure that we *understand* what we learn: every serious student should try to develop a 'feel' for the way molecules behave — for the way they are put together and especially for the 'rules of engagement' which operate when molecules meet and react.

The good news is that very much the same basic interactions control both structure and reactivity. These are stereoelectronic interactions — between electronic orbitals in three dimensions — and understanding them can give us a 'feel' for what molecules are and what they are capable of. This understanding does not have to be mathematical: my more specialised book* on the subject starts with the thought that 'a molecule's bonding electrons serve not only as its skeleton, but also as a rudimentary nervous system.'

This sort of 'feel' is especially useful for thinking about reactivity. Though vast numbers of reactions are known, they fall into a quite small number of mechanistic classes. Reading this book should convince you that these familiar mechanistic classes similarly fall into a much smaller number of classes of stereoelectronic interaction. I hope that it will also encourage you to look for the underlying stereoelectronic interactions involved in every new reaction you come across. Every reaction is a real example, and the fundamental relationships involved are the key to understanding — and predicting — new chemistry.

Stereoelectronic effects are not covered, at least by name, in most major student organic texts; though the observant reader will find evidence for them on almost every page. One general text which does treat them explicitly is: F. A. Carey and R. J. Sundberg, *Advanced Organic Chemistry, Part A, Structure and Mechanisms*, Plenum Press, New York and London, 1990.

Two specialised texts which supplement the treatment in this Primer are: P. Deslongchamps, *Stereoelectronic Effects in Organic Chemistry*, Pergamon Press, Oxford, 1983; and *A. J. Kirby, *The Anomeric Effect and Related Stereoelectronic Effects at Oxygen*, Springer–Verlag, Berlin, Heidelberg and New York, 1983.

Cambridge   A.J.K.
August 1995

Contents

1   Introduction

A molecule's bonding electrons serve not only as its skeleton, but also as a rudimentary nervous system. They relay the effects of local perturbations to other centres, more or less efficiently, depending on distance and geometry. Thus, in principle, every nucleus in a molecule can sense the presence of a strongly electronegative atom or group, or the approach of another molecule, or the changes in electron density that happen when bonds are made or broken.

In practice significant effects are quite localised, and it is these that give rise to stereoelectronic effects. They can have profound effects on molecular conformation and reaction mechanism, and are fundamental to a proper understanding of structure and reactivity.

The typical stereoelectronic effect involves an electronic interaction which stabilises a particular conformation or transition state: that is to say, it is fully expressed only when the correct geometry is achieved. Familiar examples are the inversion of configuration involved in the S_N2 reaction at sp^3-hybridised carbon, which is direct evidence for 'backside' attack by the nucleophile:

The S_N2 reaction at sp^3-hybridised carbon is discussed in Chapter 5.

and the idea that a nucleophile adding to the sp^2-hybridised carbon atom of a carbonyl group approaches from above or below, rather than in the plane defined by the group.

Nucleophilic addition to sp^2-hybridised carbon is discussed in detail in Chapter 6.

The evidence in this second case is more subtle. We cannot 'see' the way two molecules interact to form a bond, except as a theoretical exercise on the computer. But the imaginative use of evidence from series of crystal structures provides a vivid insight into what happens when a nucleophile approaches a willing carbonyl group. The most famous example is worth retelling.

Crystallographers Bürgi and Dunitz identified a remarkable structural correlation in the structures of a series of amino-ketones. The amine nitrogens are close to the carbon atoms of the carbonyl groups in these compounds, either because they are in the same molecule, or because they are brought close together by crystal packing. The closer the approach of the

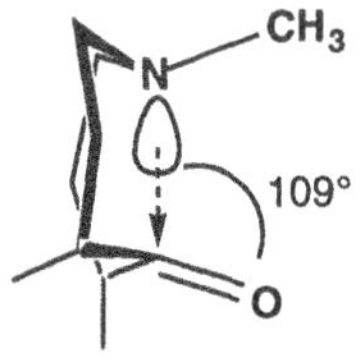

Fig. 1.1

nitrogen, the more the carbonyl carbon is seen to be displaced from the plane defined by its three substituents. This displacement is not repulsive, as might reasonably have been expected when two centres come into van der Waals contact, but evidently attractive, because the displacement is *towards* the nitrogen atom. (For example, Fig. 1.1 shows the actual geometry of this interaction in the crystal of the eight-membered ring of the alkaloid clivorine.)

In every case the amine nitrogen is observed to lie symmetrically above the ketone group, and the N···C=O angle is close to the tetrahedral angle of 109°28'. The conclusion is that this - the so-called Bürgi-Dunitz angle - defines the preferred geometry of approach for nucleophilic addition of amine nitrogen - and presumably of other nucleophiles also - to ketone carbonyl.

In practice there is a preferred geometry for practically every interaction between molecules, or parts of a molecule. The rules are relatively simple, and not difficult to work out from the principles described in Chapter 2, but they are quite fundamental to the way we think about chemistry. Every serious student should try to develop a 'feel' for the way molecules behave, and especially for the 'rules of engagement' which operate when molecules meet and react. The good news is that only a very small number of basic interactions is involved. For example, the stereoelectronic effects controlling the addition of a nucleophile to a carbonyl group also give rise to the special selectivities observed in cyclisation reactions involving the additions of nucleophiles to sp^2-carbon (**1.1** → **1.2**):

1.2 cyclises to the lactam (**1.3**) even though primary amines normally prefer to add to the C=C bonds of acrylate esters. For more details see Chapter 6.

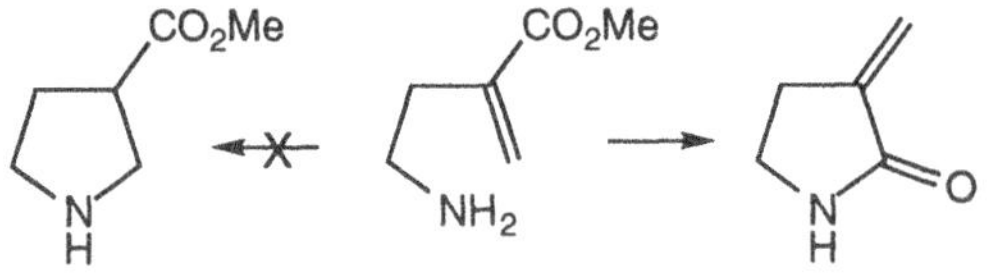

and even the unexpected conformational preference of the tetrahydropyranyl acetal (**1.3**).

The cyclic acetal exists preferentially in the conformation (**1.3a**) with the methoxy-substituent axial, even though the usual conformational preference, based on substituted cyclohexanes, is for the equatorial position. For details see Chapter 3.

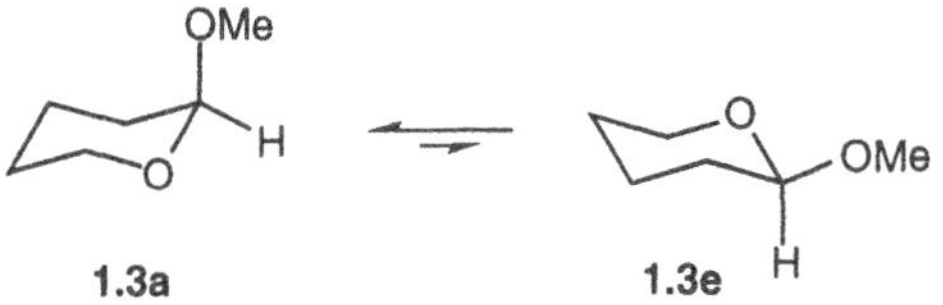

This effect is not just an academic curiosity. It controls or at least influences the conformations of key biological compounds like saccharides and phosphate diesters; and in polymeric structures the effects are multiplied hundreds or thousands of times. So - since structure and function are inextricably related in biology - an understanding of stereoelectronic effects is as essential for the thinking structural biologist as for the chemist.

Because they involve the properties of electrons in molecules stereoelectronic effects are described most appropriately in the language of orbital-orbital interactions, and the detailed discussion in the rest of this book is based on the relevant ideas and concepts introduced in Chapter 2.

Further reading
Useful general texts are described in the Preface. More specialised references appear at the end of each chapter.

2 The electronic basis of stereoelectronic effects

This chapter introduces some of the ideas on which current thinking about electrons in molecules is based. Always remember that in science, although facts (usually) remain constant, explanations may change. The ideas presented here come originally from modern theoretical chemistry, and have developed over the last twenty years into our most powerful and creative way of thinking about reactivity. The ideas have been simplified to suit the organic experimentalist, who is assumed to be non-mathematical, but has some ability to visualise objects in three-dimensions. So they are most useful, and most powerful, in a pictorial form which fits in with the way we represent molecular structures.

2.1   Orbital-orbital interactions: from AO to MO

To avoid making elementary mistakes it is important never to lose sight of what our pictorial representations mean. So we start here from the beginning, with the simplest possible theoretical picture (Fig. 2.1) of the formation of a bond between two identical atoms.

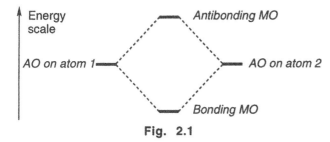

Fig. 2.1

This picture represents the interaction of two atomic orbitals (AO, identical in this case). The vertical axis is a (qualitative) energy scale and the horizontal axis gives some unspecified indication of distance on the molecular scale. The solid lines represent orbital energy levels.

The AOs might be the 1s orbitals of two hydrogen atoms or the 2p orbitals of two fluorine atoms. When they are brought together they interact to form the two new *molecular* orbitals (MO) in the centre of the diagram, one of which is lower, the other higher in energy. If the interaction is strong enough – that is to say, if the energy of the lower MO is low enough relative to the two starting AOs – electrons that were originally in the atomic orbitals will prefer to occupy the new *bonding* MO. The result will be a covalent bond – as long as the original AOs contained a total of only two electrons. If both AOs contribute two electrons, two of the four electrons involved would have

The properties of an orbital are those of an electron contained in it. It is normal practice, illogical though it may sound, to talk of 'vacant orbitals'. The properties of vacant orbitals are those calculated for electrons occupying them.

Remember that an orbital can accommodate up to two electrons. An odd number means we are dealing with radical chemistry. This situation is discussed in Chapter 8.

to be accommodated in the higher energy *antibonding* MO, cancelling out the energy gain involved in populating the bonding MO. The result is net antibonding, and a repulsion rather than an attractive interaction results.

This conclusion is the basis of the following generalisation, which has far-reaching consequences: it provides the theoretical framework not just for covalent bond formation, but also for many of our ideas about stereoelectronic effects:

> **TWO-ELECTRON INTERACTIONS ARE BONDING**
> **FOUR-ELECTRON INTERACTIONS ARE ANTIBONDING.**

The rules governing such orbital-orbital interactions are familiar. For a strong bonding/antibonding interaction:

A fourth rule is of special importance for delocalised structures. As long as Rules 1-3 are observed:

4. Any number of orbitals can interact to form an equal number of MOs

1. **The interacting orbitals must be close in energy**
2. **The interacting orbitals must overlap efficiently**
3. **The interacting orbitals must have suitable symmetry**

2.2 Interactions between molecules: HOMO and LUMO

We can use the same basic picture (*cf.* Fig. 2.1) for thinking about interactions – including reactions – between molecules. The orbitals involved may be non-bonding, like the lone pair electrons of an amine, or they can be bonding MOs – e.g. the π-electrons of an alkene – but the basic rules are the same. They need only a certain amount of interpretation in more complex situations.

Rule 1: The interacting orbitals must be close in energy

In Fig. 2.1 the interacting orbitals are identical, so clearly this first condition is met. But the great majority of reactions between molecules involve interactions between non-identical orbitals, so the resulting picture is no longer symmetrical (Fig. 2.2). The key parameter is now the difference in energy between the filled orbital of molecule 1 and the new bonding MO.

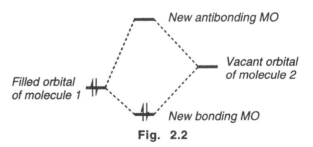

Fig. 2.2

This difference decreases as the difference in energy between the two interacting orbitals increases, until eventually significant bonding is no longer observed.

The requirement that the interacting orbitals must be close in energy means that for ionic reactions between molecules only one or two orbitals (often called the *Frontier Orbitals*) on each molecule are capable of productive bonding interactions.

Consider for example the interactions between the two identical molecules represented by the orbital diagrams shown in Fig. 2.3. Clearly the strongest – perhaps the only – significant two-electron, bonding interactions will be between the pairs of orbitals indicated by the dashed arrows. These are respectively the highest occupied and the lowest unoccupied molecular orbitals, known by their initials as the HOMO and the LUMO.

If the two molecules involved are not identical the picture is unsymmetrical (Fig. 2.4: *cf.* Fig. 2.2). Fig. 2.4 is derived from Fig. 2.3 simply by raising the energy levels of molecule 2 uniformly relative to those of molecule 1. As a result all the orbital interactions between the molecules are different, and in particular there are now two different HOMO : LUMO interactions. The energy gap between HOMO (1), of molecule 1, and LUMO (2), the lowest unoccupied orbital on molecule 2, is now too large for a significant bonding interaction. But the energies of HOMO (2) and LUMO (1) have become correspondingly closer, so these orbitals interact more strongly than before.

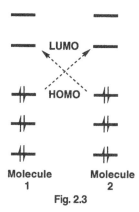

Fig. 2.3

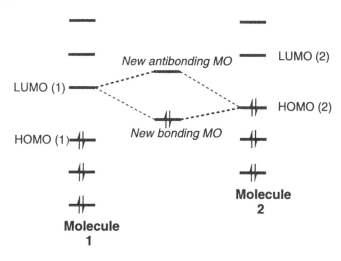

Fig. 2.4 Interacting frontier orbitals

This convergence of frontier orbital energies happens whenever the energy of the HOMO – by definition the electron donor – is raised, or the energy of the LUMO of the electron acceptor is lowered.

Rule 2: The interacting orbitals must overlap efficiently

This requirement has two related aspects. For efficient overlap the orbitals concerned must, of course, be able to approach each other closely enough; and they must be comparable in size. Orbital size and shape give an indication of local electron density, which appears in the solution of the wave equation for a given MO in the form of numerical coefficients at each atomic centre. For thinking about reactivity in typical situations very simple qualitative ideas will serve.

Fig. 2.5

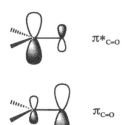

$\pi^*_{C=O}$

$\pi_{C=O}$

Fig. 2.6

Orbital coefficients are a guide to reactivity. The larger coefficient on the oxygen atom in the π-bonding orbital of the C=O group (the HOMO) tells us that electrophilic attack is expected at oxygen; while the larger coefficient of the LUMO at carbon indicates that nucleophiles will attack here.

Consider the π-system of the carbonyl group. We know that the electron-density is concentrated in two regions, above and below the nodal plane, and is polarised towards the more electronegative oxygen atom. This is reflected in a higher coefficient at oxygen for the π-bonding MO. Rather than trying to draw a convincingly distorted π-system (Fig. 2.5) it is convenient to represent this situation in terms of the two AO's involved (p-orbitals on C and O), but to give them different sizes (Fig. 2.6). The sizes represent, usually qualitatively, the orbital coefficients of interest.

The same representation of the corresponding antibonding orbital, commonly the LUMO (π^*, Fig. 2.6) for a carbonyl compound, shows that the relative sizes of the coefficients on C and O are reversed, with the larger coefficient now on carbon. When a carbonyl compound reacts with a nucleophile the decisive interaction is between the HOMO of the nucleophile and the antibonding, π^*-orbital of the C=O group as the LUMO. In terms of Fig. 2.4, Molecule 1 represents the carbonyl compound, because it has a low LUMO, while the nucleophile, which may be presumed to have a high HOMO, acts as Molecule 2. Overlap with the HOMO of the nucleophile is more efficient at the LUMO centre with the larger coefficient, rationalising the observed preference for nucleophilic addition at carbon.

Of course everyone knows that nucleophiles add to carbonyl carbon, and orbital-orbital interactions are not the only factor controlling even the kinetic products of such addition reactions. The object of this exercise is to introduce a systematic way of thinking about reactivity, which can be applied with confidence to unfamiliar situations. The next stage is to look at the symmetry of the orbital-orbital interactions involved.

Rule 3: The interacting orbitals must have suitable symmetry

Just as molecular structure can be explained in terms of the symmetries of the bonding orbitals involved, so stereoelectronic effects on reactivity can be explained in terms of the geometries of orbital-orbital interactions. And the geometries of orbital-orbital interactions depend on the symmetries of the orbitals involved.

Thus two orbitals each with axial symmetry interact most efficiently through space along a common axis. Proton transfer reactions involve substitutions at hydrogen: in frontier orbital language the non-bonded electron pair (lone pair) of the base, acting as the HOMO, interacts with the σ^*_{H-X} orbital of the acid, acting as the LUMO. The lone pair of a typical base, say NH_3, is an sp^3-hybrid orbital with an axis of symmetry, and thus defined directionality, while the $H-X$ bond of the acid (Reaction 2.1) defines the axis of symmetry of the σ^*_{H-X} orbital. Thus the most efficient overlap between the two occurs along the common axis, as shown in Fig. 2.7.

Reaction 2.1

σ^*_{H-X}

Fig. 2.7 The orbital interactions involved in Reaction 2.1.

Proton transfers between ordinary acids and bases occur along a preformed hydrogen bond, and hydrogen bonds are readily 'seen' in crystal structures because the two heavy atom centres involved are brought closer together than expected from the sum of their van der Waals radii. Though individual preferences may be overridden by rigidly fixed geometries or crystal packing forces, the clear general preference is for hydrogen bonds to form in the direction expected for the axes of the lone pair orbitals involved. For example, carboxylic acids exist almost invariably in the crystal, and often also in solution, as hydrogen-bonded dimers, with the structure shown in Fig. 2.8. The O—H bonding orbitals, and the lone pairs on the carbonyl oxygen atom of the carboxyl group, are sp^2-hybridised, and can form two linear hydrogen bonds as shown, along the axes of the orbitals concerned.

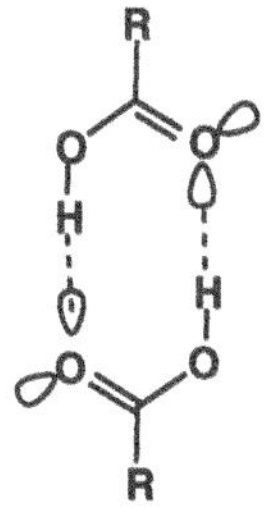

Now consider the addition of a nucleophile to a carbonyl group. We use NH$_3$, as before, so we know that the HOMO is the same as in reaction 2.1 (Fig. 2.7). The LUMO is now $\pi^*_{C=O}$, so to predict the geometry for optimal orbital overlap we must look more closely at the geometry of this orbital. The representation shown in Fig. 2.6 is good enough for most purposes, but we know that the lobes of π^*-orbitals are in fact slightly splayed out, as shown in Fig. 2.9, with their centres of electron-density beyond the two nuclei involved.

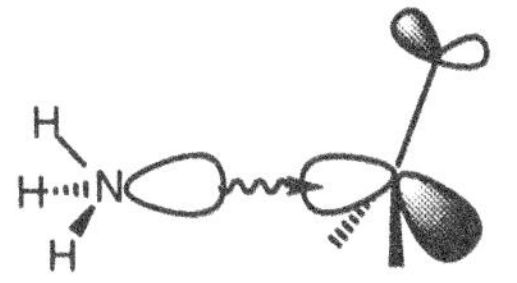

Fig. 2.9 Orbital interactions involved in Reaction 2.2

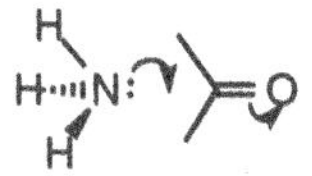

The clear prediction is that the attack of a nucleophile on a C=O group will take place from above (or below) the plane, at an angle N···C=O >90°. This is exactly the conclusion (a 'Bürgi-Dunitz angle' close to 109°: see pages 1-2 above) derived from experiment. So Fig. 2.9 shows how we visualise the geometry of this important reaction. Experimental evidence of this sort is available in only a few, specially favourable cases; but the analysis in terms of the frontier orbital interactions is always possible.

2.3   Hard and soft acids and bases

Frontier orbital (HOMO/LUMO) interactions are not the only factor involved in determining overall reactivity, though they are of course crucial in determining stereoelectronic effects. Interactions between molecules are also subject to electrostatic effects, which are not intrinsically directional, and are particularly important for reactions of charged species.

Fig. 2.8

The centres of electron-density of the separate lobes of antibonding orbitals generally lie beyond the nuclei involved in the bond concerned. This is a logical result of electrostatic repulsion: the volume between the two nuclei is already occupied by the electrons of the σ and the π-bond. (The lower energy bonding orbital is always filled first.)

Reaction 2.2

This is illustrated very clearly by the following comparisons of reactivity in simple systems:

$$HO^- \quad H_3O^+ \longrightarrow \quad 2\ H_2O \qquad \textit{very fast, exothermic}$$

$$I^- \quad H_3O^+ \longrightarrow \quad HI\ +\ H_2O \qquad \textit{relatively slow, endothermic}$$

$$I^- \quad Me\!-\!Br \longrightarrow \quad MeI\ +\ Br^- \qquad \textit{faster than hydroxide attack}$$

$$HO^- \quad Me\!-\!Br \longrightarrow \quad MeOH\ +\ Br^- \qquad \textit{relatively slow}$$

The differences in rate are substantial. Hydroxide ion removes a proton from H_3O^+ much faster than does iodide, yet iodide displaces bromide from methyl bromide faster than does hydroxide. Since the HOMO involved is the same for hydroxide or for iodide in each case, changing the LUMO cannot affect the relative strengths of the HOMO/LUMO interactions.

The accepted explanation is that the reactions of iodide ion are primarily frontier-orbital controlled, while those of hydroxide are not. Iodide has a high HOMO, which interacts strongly with low LUMOs. The HOMO of hydroxide is a lone pair on the much more strongly electronegative oxygen, so of lower energy. But for the same reason the negative charge density is more strongly attached to the oxygen nucleus, and thus *more localised* than it is in the large and diffuse 5p non-bonding orbitals of iodine. Electrostatic effects will therefore be stronger for reactions of hydroxide, and strongest for reactions with cations like H_3O^+, while the HOMO/LUMO interaction is strongest for the reaction of iodide with neutral MeBr.

Thus predicting which of two possible reactions is preferred is not entirely trivial. A nucleophile (Lewis base) may be more reactive than another towards one electrophile (Lewis acid), but less reactive towards others. To simplify the analysis the ideas involved are summarised qualitatively in terms of hard and soft (Lewis) acids and bases (HSAB). Hard acids react preferentially with hard bases, soft with soft. In HSAB terms, hydroxide is a hard nucleophile and H_3O^+ a hard electrophile, so they react preferentially with each other. The more polarisable iodide is a soft nucleophile, and prefers to react with soft electrophiles, like MeBr. The pattern of reactivity can be summarised as follows:

Reactant	HOMO	LUMO	Reacts as:
H_3O^+	-	HIGH	Hard electrophile
MeBr	-	LOW	Soft electrophile
I^-	HIGH	-	Soft nucleophile
HO^-	LOW	-	Hard nucleophile

For nucleophiles the classification as hard or soft is fairly intuitive. For electrophiles less so. The classification of some common nucleophiles and electrophiles is shown in Table 2.1: the list is artificial in that – for simplicity – it does not identify borderline cases.

Table 2.1 Typical hard and soft nucleophiles and electrophiles

Nucleophile (Lewis Base)	Electrophile (Lewis acid)
HARD: H_2O, HO^-, F^-, all oxyanions (at O)	H_3O^+, Li^+, Na^+, K^+, Mg^{++}, Al^{+++} BF_3, $AlCl_3$, AlH_3, AlR_3
SOFT: I^-, Br^-, RS^-, SCN^- $S_2O_3^{2-}$ RSH, RSR' R_3P, $(RO)_3P$ CN^- R^-, benzene	Ag^+, Pd^{++} I_2, Br_2

Ambient nucleophiles

The same sort of differences of reactivity, between different nucleophilic centres towards different types of electrophile, is also observed when the two centres concerned are in the same molecule. A group containing two different nucleophilic centres is called an *ambident nucleophile* . Familiar examples are cyanide and nitrite, and enolate anions (**2.1 – 2.3**).

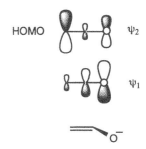

Reactions such as those shown in **2.1 – 2.3** do not usually take place in a vacuum, but in solution in a more or less polar solvent. Selective solvation of the two centres of an ambident nucleophile can have a major effect on the relative reactivity of the two centres. Hydrogen bonding solvation by protic solvents, in particular, stabilises selectively and so deactivates more highly charged centres.

In all three cases the soft and the hard centre make up part of the same delocalised system. Each nucleophilic centre reacts preferentially with a complementary electrophile. Thus each of the three anions prefers to be alkylated at the softer centre (arrows, **2.1 – 2.3**) Protonation, in contrast, is faster at the more electronegative centre, where the negative charge is concentrated.

Take, for example, the important enolate anion (reaction **2.3**). This has two π-bonding orbitals (Fig. 2.10). In the lowest energy orbital the electron density is concentrated on oxygen, as expected from its higher electronegativity. But in the HOMO the largest coefficient is on the end carbon atom, indicating that this is where frontier-orbital controlled electrophilic attack is most favourable. (The total charge, given by the sum of the coefficients for ψ_1 and ψ_2, remains greater on oxygen than on carbon.)

The reactions at the softer centres are under full frontier-orbital control, and can therefore be expected to be subject to stereoelectronic effects. But HOMO/LUMO interactions are also one of the attractive forces between reacting hard centres. So – other things being equal – the *geometries* involved will still generally be those indicated by the orbital-orbital interactions.

HOMO ψ_2

ψ_1

Fig. 2.10 The π-Bonding MOs of an enolate anion

2.4 Geometrical restrictions on orbital overlap

Many of the most interesting stereoelectronic effects on reactivity discussed in the later chapters of this book result from geometrical restrictions on orbital overlap. There is in principle no geometrical restriction on interactions between orbitals of separate molecules, apart from any imposed by steric effects. Through-space interactions between two orbitals on the same molecule, on the other hand, are subject to the constraints imposed by its structure, and this idea will be developed further in Chapter 4 (section 4.2). Through-bond interactions also clearly depend on the geometrical arrangements about the bond involved. The basic idea is a simple one, and is introduced here in the context of the structures of alkenes. It can have important effects on reactivity, and is discussed further in Chapter 4 (section 4.1) under the sub-heading of Electronic Strain.

An alkene with a twisted double bond is a reactive alkene, because π-overlap between the adjacent p-orbitals is reduced. The larger the angle of twist, the less stable is the π-bond. Thus trans-cyclooctene (**2.4**), the smallest-ring cycloalkene that is stable, is thousands of times more reactive towards electrophiles than is the unstrained cis-isomer (**2.5**). Trans-cycloheptene is not thermally stable: it is not possible to join the two allylic carbons (starred in **2.4**) by a 4-carbon chain and especially not a 3-carbon chain without twisting them too far out of the plane of the π-system.

The same effect is observed if the double bond is at a bridgehead, and is the basis of Bredt's rule: which says that double bonds are prohibited at the bridgehead positions of caged bicyclic systems. Here too the rule is relaxed when the rings become large enough to accommodate the alkene unit without prohibitive distortion. Thus **2.6** has only very transitory existence, while **2.7** can trapped reasonably comfortably and **2.8** can be isolated.

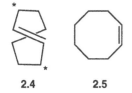

2.4   **2.5**

Note that trans-alkenes are normally *more* thermally stable than cis-alkenes, for steric reasons.

2.6   **2.7**   **2.8**

2.9

Although there is also angle strain involved with many such structures, it is the twisting of the double bond which is the key to their instability. Thus cyclopropene (**2.9** clearly suffers from enormous angle strain, but is (by definition!) planar. It is, not surprisingly, a highly reactive alkene, because much angle strain is relieved in reactions which convert the two sp^2 carbons to sp^3. But it is stable at the temperature of liquid nitrogen.

The preferred planar structure of an alkene can be distorted in several different ways. All of them weaken the π-bond, which requires the p-orbitals of the adjacent carbon atoms to be parallel to each other (**2.10a**). Simple twisting about the central C—C bond (**2.10b**) clearly reduces the efficiency of the overlap. In terms of the orbital interaction diagram on page 3 (Fig. 2.1) the new bonding MO which results is of higher energy than it would be if overlap were optimal. Since this is the HOMO for reactions of the system with electrophiles, frontier orbital interactions with a given LUMO are stronger, and reactivity is increased.

2.10a   **2.10b**   **2.10c**

The effect – like all stereoelectronic effects of this sort – is not on-off, but incremental, and depends on the angle θ between the axes of the p-orbitals. The efficiency of π-type overlap depends approximately on $\cos \theta$, falling off gently for small angles of twist (Fig. 2.11) but becoming zero at $\theta = 90°$ (**2.10**). At this point all π-bonding is lost - each p-orbital lies in the nodal plane of the other - and the two electrons that would have formed the bond must choose between the two separate orbitals. Other things being equal, the likely outcome is that they choose to separate, to form a 1,2-diradical (see Chapter 8).

Fig. 2.11

2.5   The status of lone pair electrons

Organic chemists generally think about molecular structure in terms of fixed nuclei connected by localised bonding orbitals, which fit precisely the observed geometry of the system. Non-bonding electron pairs, or lone pairs, like that of NH_3 in Fig. 2.7, are treated as orbitals much like other, bonding orbitals at the same centre: except that they are likely to be of higher energy because they are bonded to only one, rather than two, nuclei. Lone pairs are therefore the HOMOs in many systems, and of particular importance for reactivity.

2.11

Fig. 2.12. The experimentally-determined electron density map (showing bonding electrons and lone pairs only) for 9-methyladenine (**2.11**), measured in the plane of the molecule. Reproduced with permission from Eisenstein(1988), *Acta Cryst.*, **B44**, 412–26.

There is no doubt that lone pairs exist, much as represented in Fig. 2.7. Apart from the evidence from hydrogen-bonding discussed above, special crystallographic techniques reveal significant electron-density in the expected regions, consistent with high-level calculations (see Fig. 2.12); molecular mechanics calculations are inaccurate unless lone pair electron density is specifically included; and the conformational properties of amines show that the lone pair on nitrogen has steric requirements comparable with those of an N—H bond. The steric requirements of lone pairs on oxygen are also significant, but smaller, as is to be expected for a more electronegative element.

A particular and rather subtle problem arises with the two lone pairs of saturated oxygen. Amines are clearly sp^3-hybridised, with close-to-tetrahedral angles at nitrogen. The bond angles at oxygen of ethers and alcohols are also close to tetrahedral, and if the lone pairs are considered explicitly they are usually represented as two equivalent sp^3-hybrid orbitals (**2.13a**). Theoreticians prefer a representation in terms of a pair of unhybridised orbitals (**2.13b**), one with π and one with σ-symmetry, but the two pictures are mathematically equivalent, and equally valid. In either case the electron-densities of the two orbitals must add up to give the actual electron-density distribution, roughly as shown in **2.13c**). We will use sp^3-hybrid lone pairs, as in **2.13a**, which experience shows is the most useful and convenient representation for thinking about reactivity in organic molecules.

2.13a   2.13b   2.13c

2.14

Similarly the two lone pairs on sp^2-hybridised oxygen, in carbonyl compounds, nitro-groups, and so on, are represented as sp^2-hybrid orbitals (**2.14**). The simple rule is to assume that hybridisation at oxygen or nitrogen is the same as at the carbon atom they are connected to, and to draw lone pair orbitals with the appropriate geometry.

2.6   Suggestions for further reading

I. Fleming, *Frontier Orbitals and Organic Chemical Reactions*, Wiley, New York, 1977. An authoritative text that has been responsible for introducing the frontier orbital theory treatment of organic reactivity to several generations of organic chemists, and is still well worth reading today. It also gives a more detailed treatment of HSAB (pages 8-9 above), and particularly ambident nucleophiles, than is appropriate here. A second edition is expected in 1996.

A more specialised discussion of the electronic basis of stereoelectronic effects is given by Kirby (1983: for details see the Preface of this book).

3 Effects on conformation

The single most important principle of conformational analysis is that rotation about single bonds is not free, but restricted by energy barriers which depend on the substituents at the ends of each bond. Other things being equal, staggered conformations are preferred, with large, sterically demanding groups antiperiplanar - i.e. as far away from each other as possible. The preferences are easily understood in terms of steric repulsion between groups: which are in fact interactions between (usually bonding) orbitals. The general orbital interaction picture shown in Fig. 2.1 (page 3) tells us that (four-electron) interactions between filled orbitals are antibonding, and thus repulsive. These repulsive interactions are largest in eclipsed conformations (**3.1**) and minimised in staggered conformations (**3.2**). It is often convenient to discuss torsion angles in these systems using Newman projections (e.g. **3.3**).

The geometrical relationships between groups in staggered conformations, shown above relative to R*, are called *synclinal* (*sc*, in practice more often called gauche), *anticlinal* (*ac*, not much used), and *antiperiplanar* (*ap*)

3.1   **3.2**   **3.3**

The rules are based on the observed structures of innumerable saturated compounds. The few exceptions in such systems are usually special cases, where a simple explanation proves the rule. However, as one must expect, rules or explanations based on a single effect hold only as long as other effects are insignificant. And other effects do become significant, and sometimes dominant, when the C—H or C—C bonds of **3.3** are replaced by lone pairs, strongly electronegative groups, or – especially – both at the same time.

The systems of particular interest are described by the general structures **3.4**, **3.5** and **3.6**. Of these, **3.4**, which includes acetals and related compounds, is the most important. Vicinally disubstituted compounds (**3.5**) have some related properties of interest: while the conformations of systems **3.6**, which include hydrogen peroxide and hydrazine and their derivatives, fit neatly into the general picture.

3.4   **3.5**   **3.6**

Strongly electronegative atoms (X, Y) – the halogens and O, S and N, etc. – all have one or more lone pairs.

3.7   **3.8**

3.9e   **3.9a**

Note that we are dealing here with two different sorts of equilibrium. **3.9e** ⇌ **3.9a** are conformational isomers of a single compound, interconverting by ring inversion – simple rotations about single bonds. But **3.10e** is not a conformational isomer of **3.10a**, but a diastereoisomer, with one of the five stereocentres inverted. Isomerisation thus requires bond-breaking and bond-making at the centre concerned. The reaction involved is the ring-opening – ring-closing equilibration of the hemiacetal at C(1), which is known as the *anomeric centre*. The two diastereoisomers, **3.10a** and **e** (which are technically epimers in this situation), are called *anomers*.

3.1   The anomeric effect

One reason why saturated oxygen can be regarded as sp^3-hybridised (see above, page 12) is that the conformations of aliphatic ethers are much the same as those of the corresponding hydrocarbons. The preferred conformation of methyl ethyl ether, for example, is antiperiplanar, just like that of *n*-butane (**3.2** = **3.3**, R = CH$_3$); and tetrahydropyran (**3.7**) exists in the chair conformation shown, closely similar to that of cyclohexane (**3.8**).

The rules of conformational analysis tell us that the preferred form of a substituted cyclohexane has the maximum number of substituents in equatorial positions (because in this conformation the number of 1,3-diaxial interactions is minimised). The same applies to substituted tetrahydropyrans (though there are quantitative differences, as discussed below). In the 2-alkyl-substituted derivative **3.9**, for example, the equatorial conformer **3.9e** is preferred.

So we are not surprised - indeed, rather pleased - to learn that glucose, which is the most abundant naturally-occurring sugar, exists preferentially in the form (**3.10e**, β-D-glucopyranose) with all five substituents on the tetrahydropyran ring equatorial. At first sight the usual rules of conformational analysis seem to explain sugar conformations very well.

3.10e   **3.10a**

This happy state of affairs does not survive closer scrutiny. It is possible to estimate the position of conformational equilibrium for substituted cyclohexanes with high accuracy, and when the same methods are applied to the tetrahydropyran rings of pyranose sugars like glucose, clear and systematic discrepancies appear. Glucose in solution in water at 25°C does prefer the equatorial form **3.10e**, but only just. The actual equatorial:axial ratio at equilibrium is 64:36, significantly smaller than estimated, with several times more axial isomer than expected on classical conformational grounds. What is more, even in the glucose system the parent molecule turns out to be an exception: for most simple derivatives the axial anomer predominates, especially if the substituent at the anomeric position is strongly electronegative; as shown by the data in Table 3.1.

Table 3.1 Amounts of axial anomer at equilibrium for glucose derivatives

Glucose derivative	Axial isomer, %	
	R = H	36
	R = Me	67
	X = OAc	86
	X = Cl	94

This pattern is general. An electronegative substituent in the anomeric position of a pyranose, and generally in the 2-position of a tetrahydropyran ring (**3.11**), shows a definite preference for the axial position.

This unexpected preference is the anomeric effect. For OH and OR groups it is comparable with the steric preference for the equatorial position, and for the most electronegative groups it is substantially greater. All three halogen derivatives of the tribenzoate ester (**3.12**, X = F, Cl and Br: Bz = PhCO) of xylose, for example, exist in the form (**3.12a**) with the halogen in the axial position, even though this means that the three ester groups must also be axial.

3.11a   **3.11e**

3.12e   **3.12a**

3.2 Measuring the anomeric effect

The anomeric effect is obtained experimentally from measurements of the axial:equatorial ratio at equilibrium for suitable tetrahydropyran derivatives. These ratios are usually expressed as free energies, corrected for the steric preference of the group for the equatorial position, which works in the opposite direction. There are some complications with this correction, which allow some insight into the way we should think about such questions.

The relevant steric preference is that for a substituent in the 2-position of tetrahydropyran. This is not exactly the same as in a cyclohexane, but because almost all the available data refer to substituents on cyclohexane, most published estimates of anomeric effects are based on them. It would be possible to apply these data more rationally if a suitable conversion factor could be applied, and some progress has been made in this direction. The analysis is as follows.

2-Alkyl-substituted tetrahydropyrans actually show a stronger preference for the equatorial conformation **3.9e** than do the corresponding R-substituted cyclohexanes. The 1,3-diaxial interactions which destabilise the axial conformer **3.9a** are stronger in this system because C—O bonds are shorter than C—C bonds, bringing the axial H and R groups shown in **3.9a** closer than they would be in the corresponding cyclohexane **3.13a**.

The distance between carbon atoms 2 and 6 of tetrahydropyran (**3.9**, R = H) is 2.35Å, 7% less than the corresponding distance in cyclohexane. Axial substituents on these carbons are in van der Waals contact.

3.13e   **3.13a**   **3.9a**

For small alkyl substituents R the equatorial preferences for systems **3.9** are about 50% greater than for the cyclohexanes **3.13**. This figure can be used to make a rough estimate of the equatorial preference for any substituent R in the 2-position of tetrahydropyran (**3.9**) if the equatorial preference in cyclohexane (A-value) is known. It is assumed that for alkyl substituents

Equatorial preferences for substituents in cyclohexanes are called A-values, and are expressed as free energies, ($\Delta G°$) for the equilibrium **3.13e** ⇌ **3.13a**.

only steric factors are significant. For electronegative substituents X we know that the axial:equatorial ratio for the tetrahydropyran derivative **3.9** reflects both steric and anomeric effects. To separate the two the steric component has to be estimated: one suggestion is to add about 50% to the known A-value ($\Delta G^\circ_{cyclohexane}$). Some representative values for all these parameters are shown in Table 3.2. The estimated anomeric effect (final column) has not been corrected.

A-values are available in the literature for a large number of substituent groups: some examples appear in Table 3.2.

Table 3.2 Steric *vs* anomeric effects of substituents in the 2-position of tetrahydropyrans (THP, **3.11**), in kJ/mol

Substituent X	A-value	ΔG°_{steric} (THP)	ΔG° (THP) (observed)	Anomeric effect
H	0	0	0	-
CH$_3$	7	12	12	0
OH	3 - 4		−0.4	3-4
OMe	3.3	(5)	−3.7	6-7
OPh	2.7		−2.3	5
OAc	3		−2.1	5.1
SMe	4		−2.1	6.1
Cl	2.4		−7.5	10

The free energy differences ΔG° in Table 2 ((columns 3-5) are average values, for the equilibrium formation of the axial isomer (e.g. **3.11e** ⇌ **3.11a**). Thus a positive value indicates a preference for the equatorial isomer. The values are known to vary with solvent, and differences of ±1 kJ/mol are not significant.

The final column gives the magnitude of the anomeric effect for the group X, in kJ/mol. The effect increases with increasing electronegativity of the group concerned, and for the halogens is large enough to dominate conformational equilibria even for systems with several other substituents with opposite preferences (see **3.12**, above).

The examples given in Table 3.2 are chosen to cover those types of substituent which show significant anomeric effects. For groups which are only weakly electronegative any anomeric affect is small. The uncertainties involved in measuring an accurate value therefore mean that for a number of groups it is not certain whether an effect exists. This situation is fertile ground for the specialist, but for most purposes whether such groups – CN and CO$_2$Et are examples – exhibit an anomeric effect is not important.

3.3 The anomeric effect is general

The anomeric effect is a fundamental stereoelectronic effect, not limited either to acetals or to ring-systems. The conformation of the aglycone end of a glycoside, for example, shows a related conformational preference, which on inspection turns out to be basically the same effect, acting at the other end of the acetal group at the glycosidic centre. Thus the anomers of a glucoside prefer - other things being equal - conformations **3.14a** and **3.14e**. This is called the exo-anomeric effect. This preference is an important factor in the determination of the three-dimensional structure of polysaccharides, and is observed in simple tetrahydropyran acetals also (e.g. **3.15**).

The *aglycone* (literally, the part of the structure that is not sugar) is shorthand for a group (OR in **3.14a**) attached to a sugar residue.

3.14a   3.14e   3.15a   3.15e

The feature common to all these structures is a preference for a gauche (or *synclinal*, see p. 13) arrangement (**3.16**) about the central acetal C—O bonds. This is possible for both of these bonds simultaneously only in the axial geometry (**3.16** corresponds to **3.15a**). For equatorial compounds **3.15e** corresponds to **3.17**: the conformation about the exocyclic bond is gauche, but the ring enforces the antiperiplanar conformation about the endocyclic C—O bond. (Remember that a bond to an equatorial substituent is always antiperiplanar to two ring-bonds in the chair conformation of a six-membered ring).

In acyclic systems there is no overriding conformational restriction, and the stereoelectronic preference for the gauche arrangement is observed – again, other things being equal. Thus dimethoxymethane exists in the gauche, gauche conformation (**3.16**, R = CH$_3$), as do many other acyclic acetals and their derivatives.

Such conformational preferences are not limited to rotations about C—O bonds, though these are the most obvious since only a single bond is involved at the oxygen end. The simplest extension is to sulphur compounds.

3.16   3.17

Effects at sulphur

The evidence for an anomeric effect about the C—S bond is clear-cut, though quantifying it is less so. Comparable or even greater amounts of axial isomer are seen when tetrahydrothiopyran derivatives are compared with the parent, oxygen, heterocycles; increasing as expected with the electronegativity of the group (X in **3.18**) concerned.

This does not mean that the anomeric effect is actually stronger for the sulphur derivative: C—S bonds are significantly longer than C—C, so that the repulsive 1,3-diaxial interactions which disfavour axial conformations are in any case reduced in **3.18a**. This effect is greater still in dithianes (**3.19**, **3.20**) and trithianes (**3.21**). Thus trans-2,3-dichloro-1,4-dithiane, like the corresponding dioxan, exists in the crystal and in solution in the diaxial conformation (**3.19**); there is no evidence for any significant amount of equatorial isomer in solutions of the benzoyl derivative **3.21**; and even the ester group has a clear preference for the axial position in **3.20**.

3.18a   3.18e

Standard bond lengths for C—O, C—C and C—S single bonds are 1.43, 1.54 and 1.82Å, respectively. Note that, compared with C—C, the increase in length for C—S is significantly greater than the decrease on going to C—O.

3.19   3.20   3.21

Effects at nitrogen

Amine nitrogen is not electronegative enough to produce a significant anomeric effect as a substituent on a tetrahydropyran (X in **3.11**, above), but the conformational preferences of alkyl groups attached to ring-nitrogens show that similar effects are at work. For example, an N-alkyl group is axial in the predominant conformations of both **3.22** and **3.23**, and the same is true for **3.24** even when the axial group is t-butyl.

3.22 **3.23** **3.24**

This makes the conformation gauche about the C—N bond concerned (**3.25**), as observed for acetals (**3.16**, above). The range of accessible structures is however much smaller in the case of nitrogen, because N–C–X compounds are not normally stable when X is the sort of electronegative substituent that shows strong anomeric effects in O–C–X systems.

3.25

3.4 Explaining the anomeric effect

Acetals are the most important of a number of classes of compounds with two, three or four hetero-atoms on the same central carbon. In general terms, the more electronegative the substituents the more stable the system: but more and larger substituent groups mean increased steric strain. The stereoelectronic effects of interest are discussed most conveniently in the context of the familiar systems (**3.4**) with two hetero-atoms.

3.4

The second hetero-atom Y in such a molecule has a profound effect on the bond (C—X) to the first. When the two substituents are identical (X = Y) the system shows special stability: the C—Cl bonds of methylene chloride, CH_2Cl_2, or the C—O bonds of $(MeO)_2CH_2$, for example, are shorter and stronger, and thus harder to break, than those of the corresponding CH_3—X compounds. In the extreme case the C—F bonds of CF_4 are 5% shorter than those of methyl fluoride, and the decrease is monotonic with increasing numbers of fluorine atoms. The stereoelectronic origin of the effect is still under active discussion - and likely to remain so. This is the extreme case - F is the most electronegative element, its lone pairs are the most vestigial, and there is no relevant conformational evidence.

The low reactivity of methylene chloride (CH_2Cl_2), which is a useful inert solvent for organic reactions, is familiar. Symmetrical acetals are not generally thought of as inert, but in fact they react only when a proton or Lewis acid is attached to one of the oxygens, thus making the *reacting* system strongly unsymmetrical.

In the case of an acetal much more evidence is available. The substituent groups on the two oxygens adopt specific conformations, which reflect stereoelectronic as well as steric requirements, and they can be varied, together or independently, to probe further.

We have seen that the preferred structure (**3.16**) of an acetal has the gauche conformation about both central C—O bonds. This puts lone pairs antiperiplanar to both these bonds. Since we know that this conformation is not favoured sterically, we can safely conclude that the preference is stereoelectronic, and dictated by the presence of the lone pairs. Other significant evidence is that the lengths of the two central C—O bonds differ if the acetal is electronically unsymmetrical. An example is any aryl alkyl acetal (**3.26**). It is found that the C—OAr bond is longer than C—OMe, and this difference increases for Ar groups with electron-withdrawing substituents.

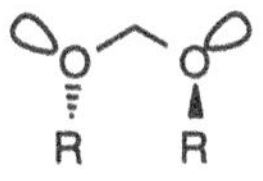

3.16 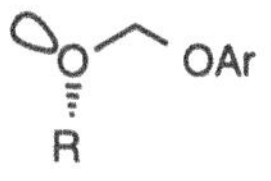   **3.26** 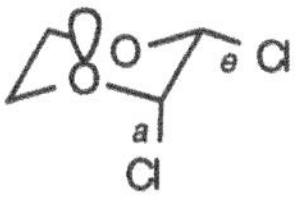   **3.27**

The same effect lengthens the axial C—Cl bond *a* in *cis*–2,3-dichloro-1,4-dioxan (**3.27**) but not the corresponding equatorial bond *e*. (The bond lengths are 1.819 and 1.781Å, respectively: a difference far greater than the errors involved in the measurement.) Here too there is a lone pair on the adjacent oxygen antiperiplanar to the bond affected.

The simplest explanation for all these effects is that there is a stabilising, two-electron interaction (expressed in terms of curved arrows as **3.28** ↔ **3.29**) between the non-bonding electron pair on oxygen and the vacant σ*-orbital of the adjacent C—O or C—Cl bond (thus n_O–σ^*_{C-O} or n_O–σ^*_{C-Cl}). The oxygen lone pairs are the HOMOs in these systems, the antibonding orbital of the σ-bond to the most electronegative substituent is the LUMO, and the antiperiplanar geometry allows optimal overlap. (As shown in Fig. 3.1 the orbitals involved have the same symmetry, and are held parallel to each other in this conformation.)

3.28   **3.29**

Fig. 3.1
n–σ* overlap, as represented
by the arrows in **3.28**.

This interaction explains all the main features associated with the anomeric effect: the changes in bond lengths at the acetal centre are predicted, since a contribution from structure **3.29** will have the effect of lengthening the C—X bond, and shortening the C—O bond, as observed. And if the effect is stabilising, as a two-electron interaction is known to be, it will be most effective when the geometry is antiperiplanar (Fig. 3.1), thus accounting for the preference for gauche conformations about acetal C—O bonds. It also predicts that acetals can be unsymmetrical purely because of their geometry. A case in point is the tricyclic system **3.30**, in which one oxygen (*a*) is axial but the second equatorial (*e*) with respect to the other ring. The conformation at the acetal centre corresponds to **3.17** (above, p. 17). The two C—O bonds differ significantly in length, with the bond to the axial oxygen (*a*) the longer: as expected, since only this C—O bond is antiperiplanar to a lone pair on the other oxygen atom.

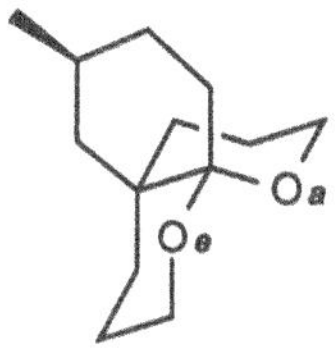

The interaction shown in Fig. 3.1 is represented in terms of curved arrows as **3.28**, corresponding to 'resonance' between the two structures **3.28** and **3.29**: neither of which represents the actual structure as well as a combination of the two. This process has been called no-bond resonance, or

3.30

'Delocalisation' in its various guises is a useful way of bridging the gap between the complex electronic structures of even quite simple molecules and our simplified picture of bonding. It is a price we pay for the convenience of describing structures in terms of localised orbitals.

sometimes negative hyperconjugation (see below). In terms of Fig. 3.1 a better general designation is σ-delocalisation.

The arrows of **3.28** represent an effect on the ground state of an acetal or related system, but the same arrows can also represent a real reaction, in which the C—X bond is broken. This is a formal connection between stereoelectronic effects on ground state structure and reactivity, and is discussed in more detail in later chapters.

This picture, involving σ-delocalisation, is the generally accepted interpretation of the anomeric and related effects, and is the one used in the rest of this book. Other rationalisations have been offered, and are discussed in the specialised texts listed at the end of this chapter. None of them has the power and simplicity of the explanation in terms of n–σ* interactions, as described above. This does not necessarily mean that this interpretation is formally or rigorously correct; simply that it is the best and most convenient currently available to the organic chemist. The most important caveat for the user in the laboratory is not that the ideas may conflict with new results of theory, but that they should be applied with due care. The effects concerned have important consequences but they are not large in energy terms, and are always in competition with other steric and stereoelectronic effects; which in many situations will be stronger.

3.5 Related effects on conformation

The basis of the anomeric effect is thus taken to be a two-electron bonding interaction between a lone pair on oxygen and the antibonding σ* orbital of a bond on an adjacent atom: this interaction being strongest when the two orbitals concerned are antiperiplanar (Fig. 3.1).

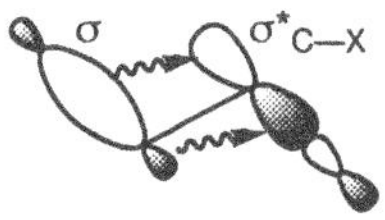

If we think of organic structures in terms of localised orbitals we must accept that such interactions, between filled and vacant orbitals, are the rule rather than the exception. Every σ-bond is also a potential donor orbital (Fig. 3.2: compare Fig. 3.1). The symmetries and the geometrical requirements for optimum overlap are the same: only the details differ.

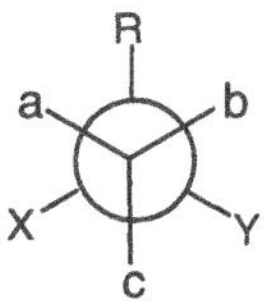

In principle there are six such interactions about every single bond in the staggered conformation, each with the correct geometry for optimal overlap. Thus in Fig. 3.3 each bonding orbital at the far end of the C—C bond, *viz.* C—R, C—X and C—Y, is antiperiplanar to a bond (C—c, C—b and C—a, respectively), and thus an antibonding orbital, at the other end: and vice versa. However, such effects become significant only when they are particularly strong. A strong HOMO-LUMO interaction requires a high HOMO and a low LUMO, and it is this combination – a lone pair in the correct position to overlap with the σ*-orbital of a bond to a strongly electronegative group – which makes the n–σ*$_{C-X}$ interaction important in controlling the conformations of acetal and related structures.

All this can be summarised in an important generalisation:

Fig. 3.2

Fig. 3.3

> THERE IS A STEREOELECTRONIC PREFERENCE FOR CONFORMATIONS IN WHICH THE BEST DONOR LONE PAIR OR BOND IS ANTIPERIPLANAR TO THE BEST ACCEPTOR BOND.

Whether the conformation which best satisfies this preference is actually observed depends on the details of the bond in the molecule concerned; but it is an essential part of any reasoned assessment of the factors involved in determining conformation.

This reasoned assessment requires an idea of the order of donor and acceptor capability of typical bonds: that is to say, the relative energy levels of the HOMOs and LUMOs concerned. The order should be intuitive for a thinking organic chemist: the best donor is, not surprisingly, a localised carbanion, the best acceptor a localised carbocation. Other donors in decreasing order of donor capability are:

$$n_n > n_O > \sigma_{C-C}, \sigma_{C-H} > \sigma_{C-X} \quad (X = N>O>S>Hal).$$

Acceptor orbitals in decreasing order of reactivity are:

$$\pi^*_{C=O} > \sigma^*_{C-Hal} > \sigma^*_{C-O} > \sigma^*_{C-N} > \sigma^*_{C-C}, \sigma^*_{C-H}$$

The gauche effect

The broad generalisation in the box above also provides a simple explanation of the gauche effect, the preference for the gauche conformation observed for systems X—C—C—Y, where X and Y are strongly electronegative groups, usually halogens. For example, 1,2-difluoroethane prefers the gauche conformation (**3.31**) in all solvents. In this conformation each C—F bond is antiperiplanar to a C—H bond, and the favourable $\sigma_{C-H}-\sigma^*_{C-F}$ interactions possible in this conformation outweigh the steric preference for the conformation with the two fluorines antiperiplanar.

3.31

The same effect is involved in the preference of the fluorine atom of 5-fluoro-1,3-dioxane (**3.32**) for the axial conformation. And is presumed to explain why it has so far proved impossible to prepare the innocuous-looking acetal **3.33**, with the trans ring-junction, from glyoxal and ethylene glycol.

3.32a   **3.32e**   **3.33**

A gauche effect is also observed for systems which have two adjacent electronegative atoms, which by definition bear lone pairs. Simple examples are H_2S_2 and H_2O_2, which have the S—H and O—H bonds almost perpendicular to each other in the Newman projection (e.g. **3.34**). Here there are no other substituents to make steric demands and the resulting conformation (as always) reflects the dynamic balance between competing steric and stereoelectronic effects. The stereoelectronic effect in this case can be interpreted in the usual way, as a preference for lone pairs to be antiperiplanar to C—H bonds, which are not very convincing as acceptor orbitals, but at least better than another lone pair. In fact it is easiest to think

3.34

3.35   **3.36**

Note that the arrows shown in **3.37Z** involve specifically the oxygen lone pair and the σ-bond of the C=O group, i.e. two orbitals in the plane of the π-system. They do not involve the π-system.

of these systems as preferring to avoid conformations with lone pairs antiperiplanar to each other. For example, the expected steric preference of the tetraalkyl hydrazine **3.35** for the diequatorial conformation is almost balanced by this stereoelectronic effect, so that the equilibrium mixture contains almost 40% of the axial-equatorial conformer **3.36**, which has the lone pairs gauche rather than antiperiplanar, as they are in **3.35**.

Effects at sp^2-centres

n-σ* and σ-σ* interactions are not limited to saturated systems. They may even be stronger when the centres involved are π-bonded, because the interatomic distances will be shorter and overlap be more efficient as a result. But conformations about double bonds will not be affected directly because they are determined by the requirements of the π-bond, compared with which the much weaker n-σ* and σ-σ* bonding are second order effects.

Effects become apparent across formal single bonds between sp^2-hybridised centres, such as the C—O-alkyl bonds of esters (**3.37**). We concluded at the end of Chapter 2 that hybridisation at oxygen or nitrogen is the same as at the carbon atom they are connected to. Thus the O-alkyl oxygen of an ester is sp^2-hybridised, as we know from the observed trigonal bond angles at this centre. The second important structural feature is the well-known preference for the Z (**3.37Z**) rather than the E-conformation (**3.37E**), which might be expected to be favoured on steric grounds (certainly for formate esters (**3.37**, R = H).

3.37Z   **3.37E**   **3.38**

The simple explanation is that **3.37Z** is stabilised by the n-σ*$_{C-O}$ interaction possible between the (sp^2-hybrid) lone pair on the O-alkyl oxygen and the σ* antibonding orbital of the C=O group (arrows in **3.37Z**). This interaction is not possible in the E-conformation, which is enforced for most lactones (**3.38**), and is thought to account for their increased reactivity compared with ordinary esters. Such relationships between conformation and reactivity are one of the main topics of the next chapter.

3.6 Suggestions for further reading

The anomeric effect has proved a popular topic in recent years for reviews and even whole books. The basics are discussed in detail in Kirby (1983: for details see the Preface to this book).

For an up to date summary which is particularly strong on recent work involving second row elements (mainly S and P) see: E. Juaristi and G. Cuevas, *The Anomeric Effect*, CRC Press, Inc., 1995. If you want to see what the specialists are currently interested in there is a recent monograph: G. R. J. Thatcher (ed.), *The Anomeric Effect and Related Stereoelectronic Effects*. ACS Symposium Series #**539**, 1993.

3.7 Problems

3.1 Satisfy yourself, using molecular models if necessary, that the gauche, gauche conformation **3.16** favoured for acetals is equivalent to **3.15a**.

3.16   **3.15a**   **3.15a'**

3.2 A second possible conformational isomer of the axial acetal **3.15a** (discussed above) is shown as **3.15a'**. Explain why this conformer is less stable than **3.15a**: and draw the third possible conformer to show why this is likely to be less favoured still.

3.3 The spiroacetal **A** has three possible conformations, **A1, 2** and **3**, but exists, apparently exclusively, in just one. Which is this likely to be?

A1   **A2**   **A3**

3.4 Explain why the acetal structure **3.33** (discussed above) is stereoelectronically uniquely unfavourable. Draw the likely conformation of the acetal **A4** which *is* formed from glyoxal and ethylene glycol.

3.33   **A4**

3.5 Explain why the conformational equilibrium **A5e** ⇌ **A5a** favours the equatorial isomer for OR = OMe, but for OR = OSO$_2$Me the axial isomer is preferred.

A5e   **A5a**

4   Effects on reactivity

This Chapter describes in general terms the various ways in which stereoelectronic effects can control reactivity, introducing ideas which are used in the following chapters.

When two independent molecules collide in solution or in the gas phase there is in principle no geometrical restriction on interactions between their orbitals, apart from any imposed by their intrinsic symmetries. In real systems there may be restrictions imposed by steric effects, but these are not relevant to the present discussion, and will not be considered further in this chapter.

Through-space interactions between two orbitals on the same molecule, on the other hand, are subject to the geometric constraints imposed by its structure. In the trivial case, we do not expect two groups to react with each other if they are on opposite faces of a rigid ring-system because the molecular geometry keeps them apart. But even when the groups concerned can get close, apparently within easy bond-forming distance, stereoelectronic effects may effectively prohibit reaction.

Through-bond interactions also lead to many important reactions. As we saw in the previous chapter, these interactions are very clearly under stereoelectronic control.

4.1 Effects through bonds

We saw in Chapter 3 (Fig. 3.2) that the n–σ* interactions that control conformation are also involved in bond-breaking (if the system is reactive enough). The same set of curved arrows describes both processes (Fig. 4.1), and the geometrical requirements for optimum overlap are the same. Thus an α-chloroether (**4.1**, X = Cl) prefers the gauche conformation shown, and reacts readily by the S_N1 mechanism because in the same conformation the lone pair on oxygen is available to stabilise the developing carbocation (**4.2**) by π-type overlap.

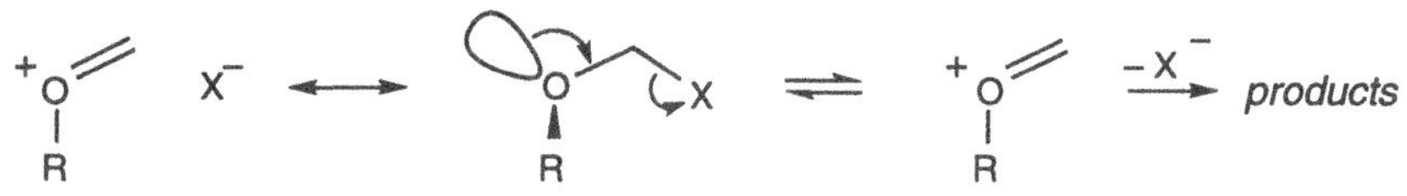

Fig. 4.1

The corresponding orbital-interaction picture for σ–σ* interactions (Fig. 3.2, p. 20) makes clear how the same factors control these interactions also.

This geometrical requirement for optimum bonding-overlap of interacting orbitals is of fundamental importance for understanding organic reactivity. As a result, electronic effects are transmitted through saturated systems most efficiently when the bonds involved are antiperiplanar. This is apparent in physical as well as chemical properties.

One of the simplest is the dependence on torsion angle of the observed three-bond coupling in ^{1}H NMR spectra between vicinal hydrogen nuclei. This coupling is well-known to be a maximum when the protons concerned are antiperiplanar (**4.3**, torsion angle $\theta = 180°$), and to follow an approximate $\cos^2\theta$ relationship (the Karplus equation), with a minimum value at $\theta = 90°$. Similarly the coupling between alkene protons is largest for the *trans* arrangement (**4.4**); while long-range (four-bond) couplings are readily observable only when the geometry about both central bonds is antiperiplanar. Thus W-coupling is observed in systems where this geometry is found in a long-lived conformation (**4.5**), as is *meta*-coupling (**4.6**) where the system is permanently fixed in the correct geometry.

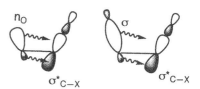

4.3   4.4   4.5   4.6

Vicinal coupling disappears, as expected, when the two C—H bonds involved are orthogonal (torsion angle = 90°), but reaches a second (lower) maximum when they are eclipsed (coplanar again, but now with a torsion angle of 0°). This is a simple indication that σ–σ* overlap is significant in this geometry also. The orbital-interaction picture (Fig. 4.2) shows qualitatively how the symmetry is correct for overlap in this situation (and closely similar for n–σ* and σ–σ* interactions), but less than optimal because the two orbitals involved are not parallel.

Fig. 4.2

Hyperconjugation

A very similar interaction between a donor orbital (lone pair or σ-bond) and a vacant p or π*-orbital accounts simply for the effect known as hyperconjugation. This accounts for the stabilisation of carbenium ions by adjacent alkyl groups which makes the t-butyl cation an energy sink while the methyl cation remains inaccessible. Once again this is a straightforward extension of a familiar idea: in this case the stabilisation of the carbenium ion **4.2** (above) by the lone pair on the adjacent oxygen. This happens because the HOMO-LUMO interaction is strongly bonding (**4.7**): as it is also when the highly electron-deficient carbocation interacts with a σ-donor orbital (**4.8**). Clearly any suitable σ-bonding orbital is a potential donor. The symmetry of the interaction shows how the same stabilising effect can operate when the acceptor is the π*-orbital of an electron-deficient π-system (C=O, etc).

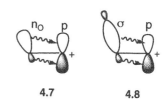

4.7   4.8

All these interactions have a direct bearing on reactivity. The ground state effect shown in **4.8** is conventionally represented as 'no-bond resonance', with a contribution from structure(s) such as **4.9**. The reality of this contribution is readily apparent in the crystal structure of the t-butyl cation (**4.10**), in which the lengths of the three (identical) C—C bonds are about half-way between standard single and double bond-lengths. We will see in Chapter 6 how the interaction-arrows shown in **4.7** and **4.8**, and those in Figs. 3.2 and 4.2, translate smoothly into reaction-arrows under the right

Remember that reactivity is determined by effects which stabilise the structures of ground states and intermediates as well as by effects on transition states.

conditions. The curved arrow in **4.10** could equally well represent the familiar reaction in which a carbocation loses a proton.

H$^+$ p

4.9   **4.10**

Negative hyperconjugation

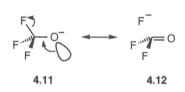

4.11   **4.12**

Negative hyperconjugation is a term that was coined to describe the stabilisation of anionic species by σ-delocalisation, involving σ*-orbitals as acceptors. An extreme case is the trifluoromethoxide anion (**4.11**), usually observed only as a highly reactive species, but stable enough in the solid state [as the $(Me_2N)_3S^+$ salt] for the crystal structure to be determined. The C—O bond in **4.11**, though nominally a single bond, has almost the same length as an ordinary C=O double bond; while the C—F bonds are significantly longer than normal. The simplest explanation is that there is strong σ-delocalisation of the non-bonding electrons on O$^-$ into the antibonding orbitals of the C—F bonds: the pattern of bond lengths is accounted for by the major contribution from the 'no-bond resonance' structure **4.12**.

It is clear from this picture that n_O–σ^*_{C-F} interaction is no different in principle from other types of σ-delocalisation we have come across (compare, for example, Fig. 4.1, above). Negative hyperconjugation is no more than a generalised anomeric effect. Thus anomeric effect, negative hyperconjugation and hyperconjugation are three terms describing in different situations what are basically the same effects.

Electronic strain

It is a truism, but one worth repeating at regular intervals, that reactivity is determined by the stabilities of ground states and intermediates as well as by effects on transition states. When the usual bonding interactions between orbitals are weakened or interrupted, for whatever reason, stabilisation energy is lost and increased reactivity is the likely result. This effect has been called electronic strain, to distinguish it from steric and angle strain, which we will not discuss here (though a case could be made that both are electronic in origin).

Steric strain is a build-up of serious non-bonded interactions, of the sort that made o-di-t-butylbenzenes difficult to make and tetra-t-butylmethane impossible. Angle strain is more obviously electronic strain: ethylene oxide, for example, is vastly more reactive towards nucleophiles than ordinary dialkyl ethers because each of the three atoms in the ring has to make bonds – that is, have orbitals interact – at unnatural angles.

Electronic strain usually results from geometrical restrictions on orbital overlap. The basic ideas involved were introduced in Section 2.3 above, where we saw how twisted alkenes show increased reactivity towards electrophilic attack. Many other systems which derive stability from π-bonding interactions show related effects when a rigid structure prohibits the planar geometry needed. Thus **4.13**, though written as an amide, does not behave like one. It is rapidly hydrolysed, under conditions where a normal amide would not react, and has a basic nitrogen (pK$_a$ 5.3). Amides are normally protonated only in strong acid, and then specifically on oxygen.

4.13'   **4.13**

Not surprisingly, the cage structure inhibits the usual amide (n_N–$\pi^*_{C=O}$) delocalisation, which would involve structure **4.13'**, with a double bond at a bridgehead. This leaves the lone pair on nitrogen available to accept a proton like any ordinary tertiary amine (albeit a weakly basic one because of the strongly electron-withdrawing group next door).

Stabilisation energies associated with the anomeric effect and with σ-delocalisation in saturated systems are relatively much smaller, and the effects of ground-state electronic strain correspondingly less dramatic. However, bond-breaking can lead directly to high energy species like carbenium ions, which derive very large stabilisation energies from adjacent σ or π-donor orbitals. The reactions of such systems can therefore still be highly sensitive to electronic strain if rigid geometries prohibit strong overlap with donor orbitals.

4.2 Effects through space

Through-space interactions between two orbitals on the same molecule are also subject to the geometric constraints imposed by its structure. When these are two-electron bonding interactions strong enough to lead to reactions such constraints can have major effects on reactivity.

The effects are most obvious for cyclisation reactions, many of which are discussed in detail in later chapters of this book. The experimental situation is conveniently summarised in a set of rules, due to Baldwin, which are set out in Table 4.1, below.

Intramolecular reactions always have to compete with the corresponding intermolecular processes. Cyclisations are prevalent because intramolecular reactions are generally faster than the same reactions between separate molecules. This is because the functional groups involved are held in close proximity. (If this sounds like no explanation at all, follow up the relevant suggestion for further reading at the end of the chapter.) Thus 4-hydroxybutyric acid (**4.14**) lactonises in the presence of acid; rather than esterifying another molecule, and giving, eventually, a polyester (**4.15**).

Intermolecular reactions, between separate molecules, in which two new bonds are formed (the Diels–Alder reaction is a familiar example) are also formally (intermolecular) cyclisation reactions. We will discuss intramolecular cyclisations almost exclusively, and call them simply 'cyclisation reactions.'

4.14   **4.15**

4.16

Cyclisation reactions are particularly favourable for the formation of stable 5- and 6-membered rings, and surprisingly easy also for the formation of (strained!) 3-membered rings. But the formation of any small ring, for

The ease of ring formation depends to some extent on the ring-forming reaction involved. A rough guide to relative rates is 5>6>3>7>4. Medium-sized (8–10-membered) rings are formed much more slowly.

example the 5-membered ring in the cyclisation of **4.14**, involves a highly structured arrangement in the rate determining step (**4.16**), which has to satisfy both the conformational demands of the system and the need for efficient orbital overlap.

Baldwin's rules

The full set of 'rules' is shown in Table 4.1. It is important to stress that these are not Rules, which have to be obeyed. (In fact one common reaction is a formal exception, as discussed in Chapter 6, page 61). Each ✔ or ✕ simply summarises a great deal of laboratory experience. ✔ means that the process concerned is generally favourable, i.e. observed: ✕ indicates that it is unfavourable, and not usually observed. Since rings of all the sizes listed can be formed if the centre concerned has the correct geometry, the cases marked ✕ are not thermodynamically unfavourable. The increased barriers to reaction are therefore kinetic, and stereoelectronic in origin.

Note: ✔ and ✕ in the Table mean that the reaction type is favourable or unfavourable, respectively: **not** allowed or forbidden. Reaction categories marked (✕) are not covered by the original rules: simple mechanisms like **4.18** may be expected to be unfavourable, but each case is of interest, for different reasons.

Table 4.1 Baldwin's Rules

Carbon-centre involved ↓	Ring-size →	3	4	5	6	7
tetrahedral, sp^3	**exo**	✔	✔	✔	✔	✔
	endo	(✕)	(✕)	✕	✕	(✕)
trigonal, sp^2	**exo**	✔	✔	✔	✔	✔
	endo	✕	✕	✕	✔	✔
digonal, sp	**exo**	✕	✕	✔	✔	✔
	endo	✔	✔	✔	✔	✔

The terminology introduced with the rules is indicated in the Table. It is widely adopted in the literature and in textbooks, and is used in the later chapters of this book. Two simple examples illustrate how it works.

At least one of the atoms involved in forming the new bond is carbon: the first column of the Table defines its hybridisation and thus geometry, as tetrahedral, trigonal or linear ('digonal'); abbreviated as **tet**, **trig** and **dig**, respectively. Column 2 specifies the geometry of the bond being broken at the carbon atom centre concerned (the same one referred to in column 1). This bond can be **exo** (outside) with respect to the ring being formed, as in **4.17**; or **endo** (inside), as in **4.18**.

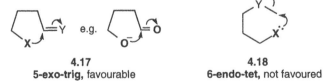

4.17
5-exo-trig, favourable

4.18
6-endo-tet, not favoured

Each reaction type is then specified, concisely, by a more-or-less self-explanatory 3-element code, as shown. Unfavourable cases, marked in the Table with **5**, are of particular interest, and are discussed in Chapters 5 and 6.

4.3 Conformation and reactivity

The remaining chapters of this book are about stereoelectronic effects on reactivity. The interpretation of differences in reactivity requires a proper understanding of the reacting system and of the mechanism of the reaction. In practice a very small number of different situations is involved. These are described in this section, which provides the basis for much of the discussion in later chapters.

It is easy enough to identify an effect on reactivity if it is large: if one isomer of a compound reacts a million times faster than another there is not much doubt that some effect is at work, and the chances are that it will be a simple matter to identify it. An example is the acid-catalysed hydration of the cyclooctenes discussed in Chapter 2 (page 10), where the electronically strained *trans*-isomer reacts much faster than the *cis*. Here the analysis is limited, implicitly, to comparing the two ground states. One is strained and thus of higher energy, so more reactive. Two-state problems are the easy ones.

In fact large effects of this sort are observed almost exclusively in rigid systems. Effects on reactivity in conformationally mobile systems are more subtle. To identify them, and especially to predict them, requires a proper analysis of the reaction concerned – which means considering activation energies. Activation energies are of course energy *differences*, between ground states and transition states. So comparing activation energies is a four-state problem: both ground state and transition state have to be considered, for both reactions.

Case 1. Isomers in equilibrium giving the same product

When comparing conformational isomers, or indeed any pair of isomers, which are rapidly interconverting, it is not possible to measure reactivity differences between them directly. Even if it is possible to start the reaction by adding pure conformer or isomer, if the interconversion is rapid relative to the reaction then the only observable reaction is that of the same equilibrium mixture $C_1 \rightleftharpoons C_2$ (see Fig. 4.3).

The fact that the lower energy, and thus more highly populated, form could react through its ground state conformation (C_1), via the high energy transition state TS_1, has little or no significance for reactivity. It has easy access to a lower energy pathway via TS_2, by way of conformer or isomer C_2. The lower energy pathway is preferred, so TS_2 determines the observed activation energy for the reaction.

In this situation the overall rate is the sum of the rates for the two forms, and the observed rate constant given by eqn (4.1):

$$k_{obs} = k_1 x_1 + k_2 x_2 \qquad (4.1)$$

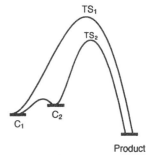

Fig. 4.3

where x_1 and x_2 are the mole fractions of C_1 and C_2. To separate the two rate constants requires an independent estimate of either k_1 or k_2.

Case 2. Isomers in equilibrium giving different products

Similar factors control the product ratio if the reactions of the two interconverting conformers or isomers give different products. In the common situation where C_1 leads specifically to P_1 (Fig. 4.4) and C_2 gives specifically P_2, a mixture of P_1 and P_2 is produced. It may at first seem surprising that the product ratio, $P_1:P_2$ bears no relation to the relative abundance of C_1 and C_2; but is determined almost exclusively by the energy difference between the two transition states TS_1^* and TS_2^* (Fig. 4.4).

Fig. 4.4 illustrates a Curtin–Hammett system, that is, one behaving according to the Curtin–Hammett principle. In its simplest form this states that the ratio of products formed from one starting material, present in two rapidly equilibrating forms, depends only on the energy difference between the two transition states, and not at all on the ground state energies of the two equilibrating forms.

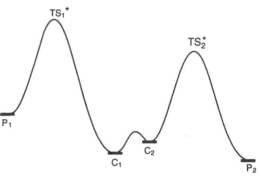

Fig. 4.4

In both these cases (1 and 2) the distinction between the two isomers is lost because of the rapid equilibrium. If the pathway which allows the ready interconversion C_1/C_2 is removed the different reactivities of the two forms can be measured without difficulty. In the case of conformational isomers this means using a modified system which is conformationally fixed: or at least strongly biased. Thus much of the classical work on conformation and reactivity was done using *trans*-decalins (e.g. **4.19**, where the substituent X is fixed in the axial position by the trans ring junction): or cyclohexane rings with t-butyl substituents. For example, it is possible to compare the rates of hydrolysis of axial and equatorial ethyl ester groups on a cyclohexane using the conformationally biased system **4.20** as the model equatorial ester.

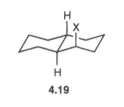

4.19

This last reaction turns out to be a good deal faster for the group in the equatorial position (**4.20** reacts some 20 times faster than the axial isomer **4.21**). Of course ring-inversion can convert **4.21** into the conformer (**4.22**) with the ester group equatorial. However, the equilibrium **4.21/4.22** is biased so strongly towards the conformation (**4.21**) with the t-butyl group equatorial – by a factor of the order of 500 – that the contribution of the unfavourable conformer **4.22** to the observed rate for the hydrolysis of **4.21** is negligible.

4.20

To sum up: there are in general two possible situations where a conformational or stereoelectronic effect on reactivity can be observed directly.

1. The barrier to equilibration ($C_1 \rightleftharpoons C_2$ in Figs. 4.2 and 4.3) is larger than the activation energy for the reaction.

2. The equilibrium is so strongly biased that the equilibrium constant is greater than the relevant difference in reactivity.

Case 3. Isomers not in equilibrium giving the same product

This third case is also a common one. We are often interested in differences in reactivity between enantiomers (in a chiral environment) or diastereoisomers where the reaction involves the formation of a common intermediate or product. E2, S_N1 and enolisation reactions are familiar examples where the same sp^2-hybridised system can be formed from such isomers.

Many of the most interesting examples are two-step reactions, and the energy profile involved (e.g. Fig. 4.5) features a high-energy intermediate (INT). To keep the discussion simple we assume that the first step of the reaction concerned is rate determining. We can therefore ignore the second step, in which the intermediate is converted to the product or products: if this step is fast enough it does not affect the rate of the overall reaction.

Fig. 4.5 may look complicated, but we have seen all its component parts before. The comparison of interest is between the reactivities of A and B. One isomer (B) has the 'correct' stereochemistry for reaction and goes directly to the intermediate. Isomer A is stereoelectronically incorrect: in its ground state form (A_1) it can react only through the high energy transition state TS_1. The stereoelectronic effect of interest is thus in principle the difference between this activation energy and that for the uncomplicated reaction of B. To separate the two in practice is usually not so simple.

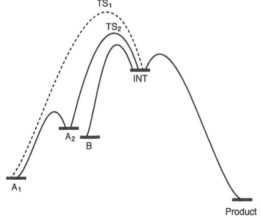

Fig. 4.5

4.23

No molecule is completely rigid. (Diamond comes close but is not a convenient test system in this context. Even adamantane (**4.23**), with the structural unit of diamond, can twist to some extent.) So – if there is sufficient thermodynamic driving force for the reaction – A has every chance of finding a way round the high energy barrier represented by TS_1. A (Fig. 4.5) is then behaving exactly in the way described above as Case 1, and the same conclusions apply. The observed rate (eqn 4.1) is the sum of the rates for the two (or more) forms, here A_1 and A_2: and TS_2 determines the observed activation energy for the reaction.

A case history

All three cases described above are nicely illustrated by a single set of experiments, which form part of the classical evidence for the preference for antiperiplanar stereochemistry in the E2 reaction. The comparison of interest (cf. Case 3) is between the axial and equatorial isomers (**4.24a** and **e**), which have t-butyl groups in the 4-position to fix the conformation, and to drive the trimethylammonium groups into the conformations shown.

No alkene is produced when the equatorial trimethylammonium compound (**4.24e**) is heated in the presence of t-butoxide in t-butanol. The E2 reaction requires the $H-C-C-N^+Me_3$ system to be at least close to antiperiplanar, but **4.24e** has two similar sterically demanding substituents with strong preferences for the equatorial position. The ground state conformation, with ring $C-C$ bonds antiperiplanar to $C-N^+Me_3$, is thus strongly favoured. Evidently reaction through any higher energy conformation is too slow to compete with the (slow) demethylation reaction, in which the base removes a methyl group in an S_N2 reaction.

4.24e → t-BuO⁻, t-BuOH, 75° → **4.26e**

4.24a → t-BuO⁻ → **4.25** (90%) + **4.26a** (10%)

4.27 → t-BuO⁻ → **4.28** (7%) + **4.29** (93%)

The Hofmann elimination of trimethylamine is, as expected, the main product from the axial isomer (**4.24a**), which has two axial hydrogens antiperiplanar to the $C-N^+Me_3$ bond. Demethylation is only a minor side-reaction in this case, indicating that elimination from **4.24a** must be at least 100 times faster than from **4.24e**. (The t-butyl and trimethylammonium groups are in fact of similar sizes, so there should be comparable amounts of the conformation shown (**4.24a**) and the isomer with the t-butyl group axial and the trimethylammonium group equatorial, in rapid equilibrium, as described in cases 1 and 2, above.)

The behaviour of trimethylammonium cyclohexane (**4.27**), is, as expected, somewhere in between. Though there is no t-butyl group to fix the conformation, the trimethylammonium group is itself large enough to enforce a strong preference for the equatorial conformation shown. It is a fair assumption, based on the result with **4.24e**, that this conformation gives only demethylation, and that the small but significant amount of elimination goes by way of the axial isomer (not shown).

This sort of evidence shows clearly that a stereoelectronic effect exists in this system, and even that it must be substantial. To quantify the effect accurately is a more difficult problem. For present purposes it is enough to note that observed differences in rate or product ratios contain information mostly about conformational preferences, and allow only minimum estimates of stereoelectronic effects.

4.4 Suggestions for further reading

The basic electronic effects are discussed in more detail in Kirby, 1983 (for details see the preface to this book).

A useful discussion of the magnitudes of $^1H - ^1H$ NMR coupling constants is given by: D. H. Williams and I. Fleming, *Spectroscopic Methods in Organic Chemistry*, McGraw-Hill, London and New York, 4th edition 1987.

Intramolecular reactivity is analysed in depth in A. J. Kirby, *Adv. Phys. Org. Chem.*, **17**, 183-278 (1980).

Conformation and reactivity is discussed in any major text on conformational analysis, e.g. E. L. Eliel, S. H. Wilen and L. Mander, *Stereochemistry of Organic Compounds*, Wiley, New York, 1994.

4.5 Problems

4.1 Explain (a) why the addition of HBr to the enamine **B1** goes only under very vigorous conditions, although enamines are normally very acid-labile; and (b) the regiochemistry of addition.

4.2. Explain why the exchange for deuterium of the proton of $HC(CF_3)_3$, catalysed by sodium methoxide in MeOD, is 10^9 times faster than the exchange reaction of the proton of HCF_3 under the same conditions.

4.3 Look up mechanisms for the Wittig and Stevens rearrangements, for which 3-endo-tet mechanisms, e.g. **B2**, can be - and once were - written.

5 Substitutions at saturated centres

This chapter deals mainly with nucleophilic substitution, though the second step of the S_N1 mechanism can be regarded as an addition reaction, and will feature also in Chapter 6. Electrophilic substitution is also discussed in this chapter; while radical substitution is covered in Chapter 8. The discussion is centred on reactions at carbon, but the mechanisms and the stereoelectronic basis of the chemistry involved are of much wider application.

Substitution, or displacement, reactions are the commonest of all organic reactions, and substitutions at sp^3-carbon were the first to be studied in depth. We know that two, conceptually quite different, mechanisms may be involved. The reaction can be concerted, with the new bond forming as the old bond breaks:

$$R-X \ + \ Y \ \rightarrow \ R-Y \ + \ X$$

or stepwise, with separate bond-breaking and bond making steps:

$$R-X \ \rightarrow \left\{ [R] \ + \ X \ \rightarrow \ [R] \ + \ Y \right\} \ \rightarrow \ R-Y$$

[R] here represents a group short of one bond. It can be a cation, as in the S_N1 mechanism for nucleophilic substitution; an anion, in which case the reaction is electrophilic substitution; or a radical.

5.1 Concerted nucleophilic substitution: the S_N2 reaction

The familiar S_N2 reaction is conventionally represented by a pair of curved arrows (Fig. 5.1), which provide a convenient and accurate indication of what is going on.

Fig. 5.1

The concerted process is accompanied by inversion of configuration at the centre where the substitution takes place, a result which is implicit in the frontier-orbital description of the reaction.

The orbital interactions involved in a simple displacement process were introduced in Chapter 2 for the transfer of a proton from an acid HX to ammonia (Fig. 5.2).

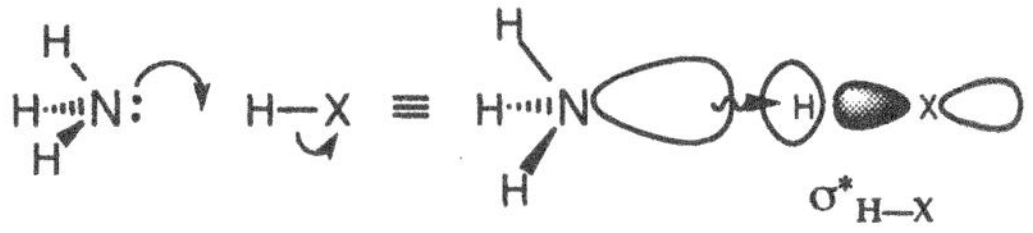

Fig. 5.2

The lone pair of NH_3 is an sp^3-hybrid orbital with an axis of symmetry, and thus defined directionality, while the H—X bond of the acid defines the axis of symmetry of the σ^*_{H-X} orbital. The most efficient overlap between the two occurs along a common axis, as shown (Fig. 5.2).

An S_N2 reaction at a saturated carbon centre involves very similar orbital interactions. The HOMO is usually a non-bonding electron-pair on the nucleophile, as it is for ammonia. The LUMO in the case of an alkyl halide, or similar system with a C—X bond to a good leaving group (X strongly electronegative) is now the antibonding orbital, σ^*_{C-X}. The most efficient HOMO/LUMO interaction is then along the common axis of the two orbitals (Fig. 5.3, top).

Fig. 5.3

Note that, purely on symmetry grounds, attack of the nucleophile might take place on X. This does not happen in this system because a carbanion is not a viable leaving group. (In frontier-orbital terms, the larger coefficient of σ^*_{C-X} is at the carbon end, so overlap is more efficient there.) But nucleophilic attack at a halogen centre is a perfectly normal reaction when it is attached to a viable leaving group, like a second halogen. Nucleophilic attack on the X—X bond involves the same interactions as the reaction with C—X shown in Fig. 5.3. The only difference is that the LUMO is now σ^*_{X-X}, and lower in energy, making the reaction with a given nucleophile much faster. Ammonia, for example, reacts sluggishly with most alkyl halides, but instantaneously with halogens.

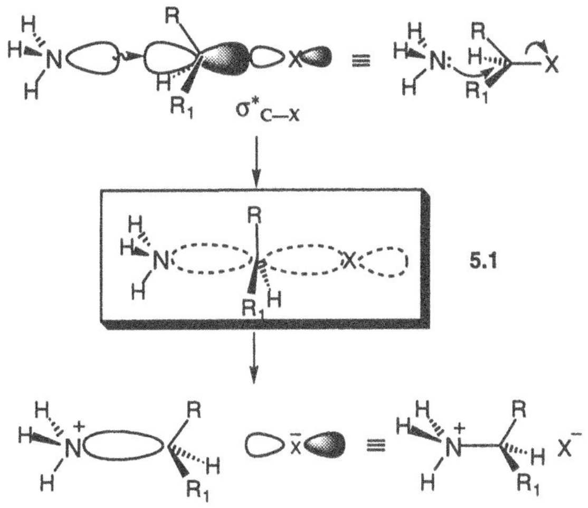

The extra dimension to displacement at sp^3-carbon is its stereochemistry. The smooth transition from the orbital system of the starting materials to that of the products, maintaining maximum overlap throughout, leads naturally to inversion of configuration. At the halfway stage there can be equal amounts of weak bonding to nucleophile and leaving group (**5.1**). If the nucleophile and leaving group happen to be the same (as in the displacement of iodide by iodide, which is a particularly easy reaction) the transition state becomes perfectly symmetrical. At this point the three 'spectator' substituents on the central carbon atom lie in a plane, and there is partial bonding to the incoming nucleophile and the leaving group.

Remember that a transition state represents the 'structure' corresponding to the point of highest energy on the reaction coordinate. But it is at the same time an energy minimum with respect to all other motion, because the reaction coordinate defines the minimum energy pathway from starting material to product.

A simple example is the equilibrium

$$PCl_4^+ + Cl^- \rightleftharpoons PCl_5$$

which allows the exchange of chloride anions at the phosphorus centre. Each exchange is formally a nucleophilic substitution.

No curved arrows are drawn for the reaction of **5.2** because the direct methyl transfer from oxygen to nucleophilic carbon does not occur.

The actual configuration of **5.1** is trigonal bipyramidal. Note that it is a transition state (an energy maximum on the reaction coordinate), and thus not a stable structure, because carbon cannot support more than four two-electron bonds. The linear arrangement of nucleophile, central carbon atom and leaving group is an essential feature: if it is not possible the reaction is not observed. Such cases are covered by Baldwin's rules (see Section 4.2 above), and discussed in the next section. Conversely, any structural feature which can stabilise the trigonal bipyramidal arrangement (**5.1**) will lower the energy of the transition state and thus favour the S_N2 reaction. This factor is the basis of a number of stereoelectronic effects on reactivity that are discussed below, in Section 5.3.

One reaction of this type that we will not discuss further is substitution at formally saturated centres other than carbon. If the element concerned comes from the third, or a higher row of the Periodic Table, it may be able to form an extra electron-pair bond using a vacant atomic orbital. This leads to a third, addition-elimination mechanism for substitution at such centres. This is similar to the familiar addition-elimination mechanism for substitution at sp^2–hybridised carbon, which is discussed in Chapter 6.

5.2 Stereoelectronic barriers to intramolecular alkyl-group transfer

The trigonal bipyramidal geometry of the transition state for the S_N2 reaction requires a linear arrangement of nucleophile, central carbon atom and leaving group, as shown in Fig. 5.3. This arrangement is not possible as part of a 6-membered ring. As a result, what may look like perfectly reasonable intramolecular alkyl-group transfers (e.g. **5.2** → **5.3**) are in fact more complicated reactions.

Labelling experiments show that methylation of the carbanion is in fact an intermolecular reaction: as we should now expect, because the intramolecular reaction would be 6-endo-trig, a class disfavoured according to Baldwin's rules (see Section 4.2, page 28). Thus in an experiment where the reaction was stopped before it was complete, by neutralising the carbanion, a mixture of three products **5.3 – 5.5** was obtained), showing clearly that **5.3** is not the initial product of reaction.

5.2 →

5.3   5.4   5.5

5.6

5.7

As usual, the stereoelectronic barrier is not absolute: it is not necessary for the N····C····X angle at C to be exactly 180°, and borderline cases are to be expected. Most of these are of little synthetic interest, and endo-tet reactions involving medium-sized rings face the additional disadvantage of conformational strain. Thus methyl transfer to nitrogen from the sulphonate ester group of **5.6**, to give the zwitterion **5.7**, is still an intermolecular process.

Reactions involving the epoxide ring

Three-membered rings stretch all the rules. They are easily formed, despite the angle-strain associated with internal angles close to 60°, so 3-exo-tet is evidently a favourable process: even though at first sight the geometry of the S_N2-type transition state (e.g. for epoxide formation, **5.8**) looks worse than the one rejected by **5.6**.

5.8   5.9   5.10

A helpful way of thinking about the stereoelectronic arrangements in this reaction is to visualise the HOMO/LUMO interaction between $\sigma*_{C-Cl}$ and the centre of electron density of the oxygen lone pair (**5.10**), rather than taking literally the angles defined by the reacting nuclei. This is legitimate in this special situation because σ-bonds in three-membered rings are considered to be bent, with centres of electron-density outside the lines connecting the nuclei (see below, page 74).

There are strict limits on the degree of tolerance allowed, as shown by the conformational control of epoxide opening and formation in six-membered rings (compare the reactions of **5.11** and **5.12**). The in-plane arrangements (*cf.* **5.10**) may seem to bend the rules, but the requirement that all the orbitals (and therefore all the atoms) involved must lie in the same plane is strict. The correct trans-coplanar geometry is easily available only when oxygen nucleophile and leaving group are both axial (**5.11**). The more difficult this geometry is to attain, the more likely that some other (or no) reaction will be observed. Thus **5.12** does eventually give epoxide, but the reaction is thousands of times slower, and presumably goes through a high energy twist-conformation to get the oxygen and bromine centres more or less coplanar.

The trans-ring junction in **5.11** and
5.12 (the bracket represents a second,
trans-fused 6-membered ring)
prevents ring inversion. But the free
end of the ring retains a good deal of
conformational flexibility.

The most important consequence is for the reverse reaction, when an epoxide is opened by a nucleophile. A nucleophile attacking epoxide **5.13** has a choice of two similar secondary centres, but attack at one of them must go through the high energy conformation, and thus transition state, involved in the slow cyclisation of **5.12**. The observed ring-opening reaction thus gives specifically the diaxial product corresponding to **5.11**. Here the combination of stereoelectronic and conformational preferences determines which of two different products is formed.

When an epoxide is opened by an intramolecular nucleophile Baldwin's classifications are less clear-cut. To take just one example: the unique geometry of the three-membered ring allows the 5-exo-tet reaction (**5.14**), but is reasonable also (**5.15**) for the formation of a six-membered ring. It can be argued that the epoxide oxygen is just as much an exo as an endo-leaving group with respect to the six-membered ring (compare reactions **5.15** and **5.16**). This is one of those points where fine distinctions can become a matter of semantics. The recommended avoiding action is to go back to the basic stereoelectronics and examine each situation on its merits. The important consequence is that competition between 5-exo and 6-endo-ring opening, for example, is much more equal for epoxides than for other leaving groups. Both are possible, and the balance can be changed by suitable patterns of substituents on the epoxide carbon atoms.

5.3 Stabilisation of the S_N2 transition state by delocalisation

The transition state represented by **5.1** (Fig. 5.3) is a high energy species, and very sensitive to local structural features which may stabilise or destabilise it. The two partial bonds, derived by mixing a filled and a vacant orbital, can be regarded as having to some extent the character of both. This means that stabilising delocalisation should be available from both electron donor and vacant orbitals, and this becomes significant if good donors or acceptors are involved.

The strongest interactions are with π-donors and acceptors. The three most important examples are summarised in Fig. 5.4, and discussed in turn below. The symmetry allows efficient overlap with any p or π-type orbital: the effect observed depends on its energy level as either donor or acceptor.

There is no doubt, for example, that an adjacent carbonyl group (Fig. 5.4, **5.1** + $\pi^*_{C=O}$) will act exclusively as a π-acceptor. Conversely an adjacent oxygen or nitrogen acts as expected as an electron-pair donor (e.g. **5.1** + n_O). But the HOMO and LUMO associated with an adjacent π-system (allyl or benzyl, shown in Fig. 5.4 specifically as **5.1** + $\pi_{C=C}$) can vary a great deal, depending on substituents. The S_N2 transition state can thus derive stabilisation by overlap with either.

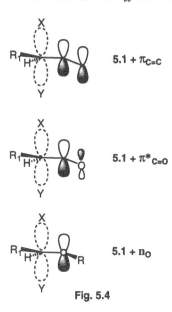

Fig. 5.4

S_N2 reactions at allyl and benzyl centres

The stabilisation of the S_N2 transition state available from bonding overlap with π-donor orbitals (Fig. 5.4, **5.1** + $\pi_{C=C}$) provides a simple explanation of the well-known high reactivity of allyl (**5.17**) and benzyl derivatives (**5.18**) in S_N2 reactions, compared with the corresponding saturated aliphatic compounds. Thus dimethylaniline is alkylated by allyl bromide over 100 times faster than by ethyl bromide; and benzylic halides react faster still when compared with the corresponding ethyl derivatives.

5.17

5.18

There is evidence in some cases that an adjacent $\pi_{C=C}$-system can also act as a π-acceptor. A double bond conjugated to a C=O group, for example, acts exclusively as a π-acceptor. More subtle are a number of S_N2 reactions of substituted benzyl derivatives (**5.18**) which are accelerated by both electron-withdrawing and electron-donating substituents Y. The likely explanation is that, for Y = H, π–donor and π-acceptor stabilisation are roughly in balance. The effect of the substituent is then to increase either one more than it decreases the other.

The importance of π-stabilisation is brought home very clearly by what happens when this delocalisation is not available. The sulphonium compound **5.19** is debenzylated smoothly by thiocyanate, as expected: both the benzylic centre and the sulphur atom of the thiocyanate anion are well-known to show high reactivity in the S_N2 reaction. Attack on the ethyl group is not observed, and must be much slower than attack at benzylic carbon.

Allyl and benzyl derivatives with good leaving groups also undergo rapid S_N1 reactions. The source of the stabilisation is the same π-donor system as shown in **5.1** + $\pi_{C=C}$, but now the acceptor is the developing p-orbital of the carbocation, potentially a more powerful electron–sink (lower energy LUMO).

5.19   **5.20**   **5.21**

Just how much slower is suggested by the result with the cyclic sulphonium compound **5.20**, which has the same arrangement, of an ethyl group and two benzylic carbons on the sulphonium sulphur. The reaction is now 8000 times slower, under the same conditions. Moreover, the main product is now the heterocycle **5.21** (80%), produced by attack on the ethyl group: only 20% of the reaction of thiocyanate with **5.20** takes place at benzylic carbon. The competition for the nucleophile between the ethyl and benzylic centres is now more closely balanced, because in **5.20** the benzylic C—S bonds are held by the five-membered ring more or less in the plane of the π-system of the aromatic ring. As a result the π-delocalisation that makes a normal benzylic centre so reactive (**5.1** + $\pi_{C=C}$ in terms of Fig. 5.4) is not available, and the benzylic CH_2 groups are no more reactive than any other primary centre.

S_N2 reactions next to carbonyl groups

The discussion in this section is limited to reactions next to C=O, but the conclusions apply equally to other electron-deficient systems with π-symmetry, such as C=N, C=N$^+$ and C≡N.

When the adjacent π-system is a carbonyl group the only significant delocalisation of the S_N2 transition state is with the antibonding π* orbital. Thus S_N1 reactivity is not observed, but concerted reactions with nucleophiles may be strongly accelerated. Depending on the nucleophile and the conditions reactions may be even faster than the corresponding reactions at benzylic centres. The displacement of chloride by iodide from phenacyl chloride, $PhCOCH_2Cl$, for example, is 10^5 times faster than from a simple primary alkyl chloride.

The interpretation in terms of delocalisation into $\pi^*_{C=O}$ ((Fig. 5.4) is supported in this case also by experiments with systems where overlap is prohibited by the geometry. The simplest to interpret is a comparison between the opening of the ring of **5.22** by iodide and the corresponding intermolecular displacement, of effectively the same very good leaving group from the phenacyl system (**5.23**).

5.22   **5.23**

The reaction of the open-chain compound **5.23** is several thousand times faster at low temperatures than that of the ring-opening of **5.22**. The difference is entirely accounted for by a less favourable enthalpy of activation, as expected for an electronic effect. The explanation is that the C—O bond to the leaving group (starred in **5.22**) is held in the plane of the carbonyl group by the five–membered ring: which is made up – apart from the CH$_2$ group under attack – of four sp^2-hybridised atoms, and thus unable to pucker even to the small extent possible for a normal ring of this size. The transition state for the S$_N$2 reaction of **5.23** can take advantage of the normal stabilisation available from delocalisation into $\pi^*_{C=O}$.

Reactions next to hetero-atoms

We have already seen (Section 3.4) that the second heteroatom in a system X—C—Y has a profound effect on the bond to the first. When the system is symmetrical (X = Y) the two C—X bonds are shortened and strengthened. This is explained, at least in part, by the n–σ^*_{C-X} interactions of the anomeric effect working in both directions. But when the system is unsymmetrical, as in an α-chloroether, the n$_O$–σ^*_{C-Cl} interaction is much stronger than n$_{Cl}$–σ^*_{C-O}, and can lead to cleavage of the C—Cl bond (**5.24**: see Section 4.1).

5.24

This is an S$_N$1 reaction, and is thus particularly rapid for an α-haloether. But so too is the S$_N$2 reaction (see the Table in Problem 5.3, below). This is the expected result of overlap between a lone pair on oxygen and the partial bonding system of the S$_N$2 transition state (**5.1** + **n$_O$**) suggested above in Fig. 5.4. (The corresponding α-halo-amine structure is not normally stable enough to exist because the lone pair of amine nitrogen is a better donor still.)

The effect can be reduced either (a) electronically, or (b) by preventing overlap by fixing the geometry. Thus (a) α-chloroalkyl esters, RCOOCH$_2$Cl, are hydrolysed by normal ester hydrolysis mechanisms (attack at C=O); and the α-bromoether **5.25** is very unreactive in S$_N$2 (or S$_N$1) reactions: and (b) the bicyclic α-chloroether **5.26** is actually less reactive than the corresponding chloride **5.27**.

5.25

Of course these bicyclic systems cannot react by the S$_N$2 mechanism because backside attack is prevented by the ring structure. But the message from Fig. 5.4 is clear: the features which stabilise a developing carbocation, and thus support high reactivity in the S$_N$1 reaction, will stabilise the S$_N$2 transition state also. So that any system which reacts readily in the S$_N$1 mode is potentially reactive by the S$_N$2 route also. In **5.26** and **5.27**, and – albeit less dramatically – at other tertiary centres, this potential reactivity is masked by steric effects.

5.26   **5.27**

5.4 The S$_N$1 reaction

The factors which stabilise the transition state for the S$_N$2 reaction are even more effective in stabilising a carbocation. The symmetry is in principle the same for delocalisation of the positive charge by adjacent donor orbitals (compare Figs. 5.4 and 5.5). So two of the three most important interactions in Fig. 5.4 – with adjacent π-systems and lone pair orbitals – also enhance reactivity in the S$_N$1 reaction. In contrast, adjacent C=O can interact efficiently only through its π^* orbital: since no bonding overlap is possible between two vacant orbitals, an adjacent C=O group generally inhibits S$_N$1 reactivity.

The HOMO of the carbonyl group ($\pi_{C=O}$) is too low in energy to interact productively even with the vacant p-orbital of a carbocation. Though even this interaction has to be considered under extreme conditions. There is recent evidence that suggests that the HOMO of the C≡N group can act as a π-donor in such a situation.

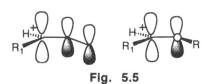

Fig. 5.5

Two general points need to be borne in mind when reactivity in the S$_N$1 reaction is discussed. First, the transition state is not the carbocation, as drawn for example in Fig. 5.5, but a species with a partial bond to the departing leaving group. Since carbocations are in most cases high energy intermediates the transition states leading to them are expected to be close to them in energy, and therefore in structure. (A statement of the Hammond Postulate.) It is therefore a convenient simplification – *but a simplification nonetheless* – to discuss reactivity in terms of carbocation stability.

The Hammond Postulate can be simply summarised: adjacent states on a reaction coordinate that are close in energy are close in structure also. The effect is illustrated by the energy profiles of Fig. 4.5, page 31. The higher the energy of the intermediate (INT) the closer it will be to the various adjacent TS.

The second point is more subtle. The transition state for the S$_N$2 reaction (**5.1**, p. 35) is more or less infinitely variable. Such a high energy species will accept stabilisation from any available source. For reactions in solution the available sources are the orbitals of the starting material and the product – i.e. the carbon skeleton, the nucleophile and the leaving group – and the solvent. Since a purely dissociative S$_N$1 reaction generates a pair of ions it is likely to be run in a polar, ionising solvent; which is itself a potential nucleophile. (In a solvolysis reaction it is both solvent and nucleophile, by definition.) The result is that all S$_N$2 and many S$_N$1 reactions go through related transition states: which can be represented in general terms as in Fig. 5.6 (i.e. no different in principle from **5.1** above).

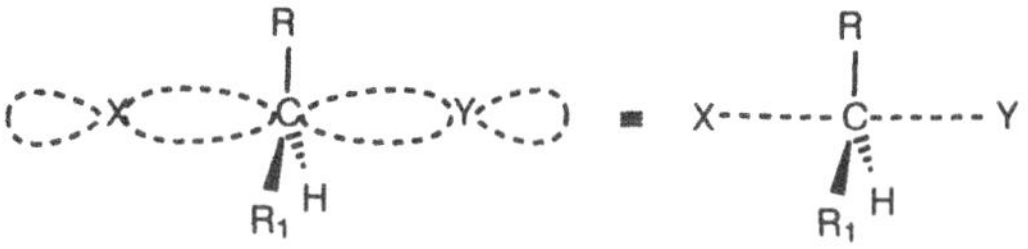

Fig. 5.6

Even for well-established S$_N$2 reactions a range of structures is possible. At one extreme are ordinary, 'tight' transition state structures, where there is strong bonding to both X and Y. At the opposite extreme are 'loose,' sometimes called 'exploded' (not a good name for a theory) transition states, with little bonding to either X or Y, and an accumulation of positive charge at the carbon centre.

At this point the picture merges with the range of transition states believed to be involved in many S_N1 reactions; where solvent takes the place of the incoming nucleophile, and thus assists the reaction, without appearing in the rate law. This is thought to happen in many reactions once thought to go via the unassisted S_N1 pathway. Two useful generalisations, based on the stability of the potential carbocation, also go some way to explaining this convergence of transition state structure: the higher the energy of the carbocation as a full intermediate in an S_N1 reaction, the more likely it will be that some nucleophilic assistance is involved: alternatively – now considering the reaction in terms of an S_N2 mechanism – the more stable is the same (potential) intermediate carbocation the looser the transition state is likely to be. The result in both cases is the so-called loose transition state.

Complicated though this situation may appear, the consequences for stereoelectronic effects are simple. The transition state for an S_N1 reaction (or for an S_N2 reaction with a loose transition state) develops significant positive charge at the central carbon atom. This has to be delocalised, or the reaction will not happen. Any available π-donor will stabilise a carbocation and thus assist the S_N1 reaction, as long as efficient orbital overlap is possible.

The new factor in reactions at the S_N1 end of the spectrum is the importance of stabilisation by σ-donors. Because carbocations are high-energy species, and thus have particularly low-lying LUMOs, they can engage in significant bonding interactions with donor orbitals which are too weak to stabilise the typical S_N2 transition state significantly.

σ-Delocalisation in S_N1 reactions

The idea of σ-delocalisation has already been introduced under the heading of hyperconjugation, in section 4.1. Bonding interactions (**5.28**) with the σ-bonding orbitals of adjacent C—H bonds are the basis of the stability of tertiary alkyl cations such as t-butyl (**5.29**)

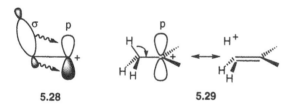

5.28   **5.29**

Any σ-bond in the correct position and with the correct orientation is a potential donor, but σ_{C-H} is by far the most important: partly because it is ubiquitous, but also because the interaction is very sensitive to the energy of the σ-orbital involved, and σ_{C-H} is one of the best donor orbitals.

A C—X bond is not usually a useful donor if X is more electronegative than H. C—C can be involved (sometimes with far-reaching consequences: see the following section), because sp^3-C is little (if at all) more electronegative than hydrogen. However, where X is *less* electronegative than H the σ-donor effect (Fig. 5.7) will be stronger than for C—H, and may come to control the chemistry.

Fig. 5.7

The case where X is Si is of particular interest. The stabilisation of a carbocation by a silyl group in the β-position (e.g. **5.31**) is the basis of a great deal of recent organosilicon chemistry. To take one example important in synthesis, the attack of an electrophile on an allylsilane (**5.30**) takes place predictably at the end of the double bond remote from silicon (at least so long as the geometry of the system allows the overlap shown in Fig. 5.7).

Note: the silyl group is lost from **5.31** in much the same way as a proton would be. When the group lost is not H$^+$ but a carbocation this step is called a fragmentation (see Chapter 7).

$$R_3Si \diagup\diagdown\diagup \xrightarrow{E^+} R_3Si \diagup\diagup\diagdown E \longrightarrow \diagup\diagdown\diagup E$$

5.30   **5.31**

This represents more powerful σ-donation than we have seen before. Even where there is competition from π-stabilisation reaction may still be directed by silicon, as shown by the example in Fig. 5.8.

$$Me_3Si\diagdown\diagup\diagdown Ph + \diagup\diagdown\diagup_O \xrightarrow[CH_2Cl_2]{TiCl_4} \diagup Ph$$

Fig. 5.8

σ-Delocalisation involving C—C bonds.

To be a good σ-donor a C—C bonding orbital needs to be of particularly high energy. (The implicit benchmark is always σ$_{C-H}$.) The strained C—C bonds of cyclopropane rings are particularly suitable, and well-known to be available for delocalisation even with ordinary electron-deficient π-systems. Thus cyclopropyl ketones (e.g. **5.32**) show low carbonyl stretching frequencies in the infra-red. Similar effects on the UV spectra of related conjugated systems are comparable with, or even greater than the effects of extra double bonds; and indicate clearly that the cyclopropane ring is involved in delocalisation.

Where the σ-donor orbital stabilising an existing or developing carbocation is a C—C bond the integrity of the carbon skeleton is at risk. Such an interaction may lead to a rearranged structure, or even C—C cleavage. This sort of reaction is discussed specifically in Chapter 7. We concentrate here on cases where stabilisation of the existing structure is the most important effect.

Another result of the high energy of the C—C bonding orbitals of cyclopropanes is that they can undergo addition reactions. This aspect is discussed in Chapter 6.

These effects depend characteristically on geometry, as expected for stereoelectronic effects. They are strongest in the so-called *bisected* conformation (**5.33**), and minimal in the *perpendicular* form **5.34**. (The terminology refers to the relationship between the plane of the cyclopropane ring and the plane defined by the substituents on the adjacent sp^2-hybridised centre.) In the bisected conformation there is optimal overlap between the antibonding orbital (here π*$_{C=O}$) of the π-system and two of the three σ-bonds of the ring.

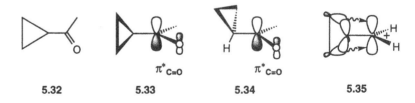

5.32   **5.33**   **5.34**   **5.35**

The same effect stabilises the cyclopropylmethyl carbenium ion (**5.35**) and derivatives; these are actually more stable than the corresponding benzylic cations. Crystal structures confirm that the preferred geometry is that shown. The delocalisation must be symmetrical, and is represented most conveniently by the overlap shown in **5.35**, in which two of the bent bonding orbitals of the ring are involved in a bonding interaction with the vacant p-orbital of the carbocation.

Reactions involving these cations as intermediates show clear evidence for σ-delocalisation. The products from the hydrolysis of cyclopropylmethyl chloride (**5.36**), for example, are a mixture of the three isomeric alcohols **5.38** – **5.40**.

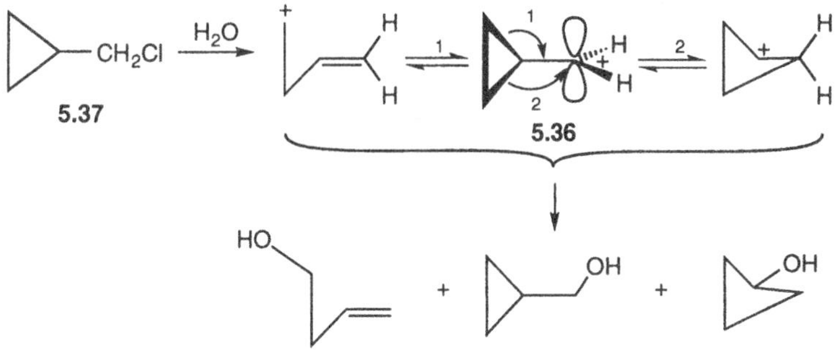

Fig. 5.9

σ-Delocalisation in situations like this took many years to become accepted, and cations like **5.36** were christened 'non-classical.' The area is a rich source of fascinating stereoelectronic effects, but far too extensive to go into in any detail here. However, the familiar generalisation which is the key to understanding almost all stereoelectronic effects of this sort is a reliable guide:

> **WHERE THE GEOMETRY ALLOWS, OR IMPOSES, GOOD OVERLAP BETWEEN A σ_{C-C} BONDING ORBITAL AND THE VACANT p-ORBITAL OF A HIGH ENERGY CARBOCATION, σ-DELOCALISATION IS TO BE EXPECTED.**

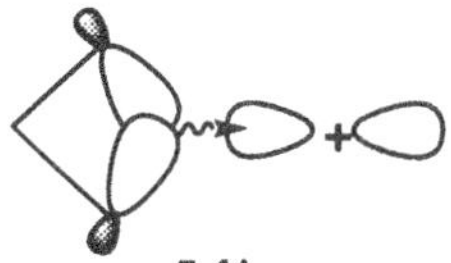

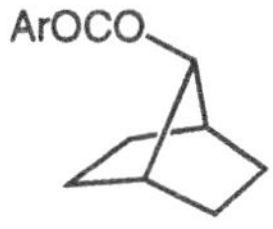

One final example involves the σ-bonds of cyclopropane in a quite different way. Through-space interactions generally have different geometrical preferences from the through-bond interactions responsible for the σ-delocalisation we have discussed so far. The most efficient interaction between a σ-bond and an orbital with p-symmetry which are not directly connected by a σ-bond is along the axis of the p-orbital, and from the centre of electron-density of the σ−bonding orbital (**5.41**). This interaction is involved in through-space neighbouring group participation by the cyclopropane ring. For example, it is the basis of the extraordinarily fast solvolyses of systems like **5.42**. The ester (Ar = 4-nitrobenzoate ester) is solvolysed almost 10^{14} times faster than the corresponding ester (**5.43**) (which has an ordinary σ-bond in the position where a nucleophile could assist the departure of the leaving carboxylate group), and gives the rearranged alcohol **5.45** as one of the products. The electron-rich σ-bond of the cyclopropane is perfectly placed to stabilise the cation formed. The delocalisation involves two electrons but three centres, and is represented by the dashed triangle shown for **5.44**.

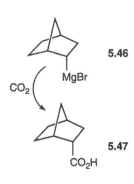

| **5.42** | **5.44** | **5.45** |

5.5 Electrophilic substitution at saturated centres

In electrophilic substitution a new bond is made to an electrophile, so the relevant frontier orbitals are the LUMO of the electrophile and the HOMO of the reactant of interest. In the concerted mechanism at a saturated centre the HOMO is by definition a σ-bonding orbital, so compounds with electron-rich σ-bonds, as discussed in the previous section, will show particular reactivity. But in principle any σ-bond can react if the LUMO is attractive enough, and under super-acid conditions even C—H bonds are attacked by protons. In the extreme case the protons of CH_4 have been observed to exchange with deuterons, and the stable t-butyl cation is formed from isobutane.

$$Me_3C\!-\!H \;+\; FSO_3H.SbF_5 \;\rightarrow\; Me_3C^+ + FSO_3^-.SbF_5 + H_2$$

In this reaction (one of several observed under the conditions) the electrophile ends up bound to H. Hydrogen is formally less electronegative than carbon, and the t-butyl is a more stable cation than an unsolvated proton, so we can deduce that electrophilic attack on C—H bonds is not likely to be very useful in synthesis. What is needed for clean electrophilic substitution at carbon is a C—X bond where X^+ will give a stable cation.

$$R\!-\!X \;+\; Y^+ \quad \rightarrow \quad R\!-\!Y \;+\; X^+$$

The obvious choice is a metal (i.e. X = M), and by far the most important type of electrophilic substitution in synthesis involves organometallic compounds. These are generally not ideal subjects for mechanistic work, but at least the stereochemistry of the reaction is well-defined. Retention of configuration is the general rule, as seen, for example, in the carboxylation of the Grignard reagent **5.46** to give the carboxylic acid **5.47**.

However, this preference is not nearly as strong as the preference for inversion in the S_N2 reaction, and inversion at the carbon centre can be observed under the right conditions. A carefully designed example is the reaction of the sterically hindered tin compound **5.48** with bromine.

$$(\text{t-BuCH}_2)_3\text{Sn}\!-\!\overset{\text{Et}}{\underset{\text{Me}}{\diagdown}}\text{H} \;\xrightarrow{\;\text{Br}_2\;}\; (\text{t-BuH}_2\text{C})_3\text{Sn}\!-\!\text{Br} \;+\; \text{H}\!-\!\overset{\text{Et}}{\underset{\text{Me}}{\diagdown}}\text{Br}$$

5.48

All these results are simply interpreted in terms of the frontier orbitals involved (Fig. 5.10). The HOMO is the highest energy σ-bonding orbital. This will be the one to the least electronegative centre, which we will call M. The LUMO will normally be a σ*-orbital or a vacant p-orbital, and thus have an axis of symmetry. The most efficient overlap is between the LUMO, along its axis, with the centre of highest electron-density of the HOMO. This is in the central region of the σ-bond, nominally perpendicular to the bond, but at an angle determined by the steric requirements of the system. If the carbon centre is more sterically demanding the interaction is represented most simply as shown in Fig. 5.10. Retention of stereochemistry results naturally from front-side attack.

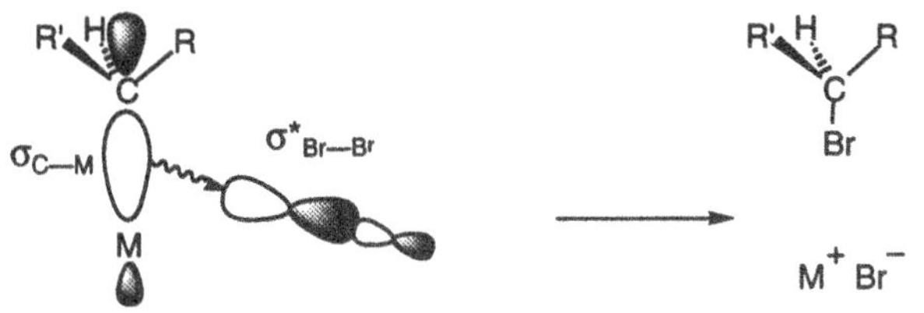

Fig. 5.10

An alternative mode of attack – along the common axis of the HOMO and the LUMO – also has the correct symmetry. Here the LUMO (a vacant p-orbital in Fig. 5.11) engages the 'tail' of the σ-bonding orbital, which has significant electron-density in a bond polarised towards carbon. The geometry of this interaction parallels that for the S_N2 reaction, and inversion of configuration would be the logical result (Fig. 5.11).

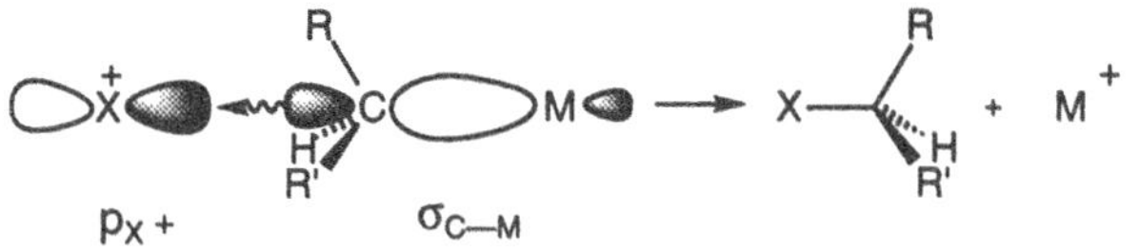

Fig. 5.11

Electrophilic attack on electron-rich C—C σ-bonds, such as those found in cyclopropanes, involves the same two interactions (Figs. 5.10 and 5.11). However, the result is generally addition rather than substitution, so is dealt with in the next chapter. The difference arises because a carbocation is not normally as stable as the M^+ used in typical S_E2 reactions, and is not released into solution.

The logical exceptions are systems like cyclopropanols (**5.49**), where the developing carbocation is effectively stabilised, in this case by π-donation from the adjacent oxygen. This is an interesting system, which is hydrolysed with either retention or inversion, depending on the conditions. When the electrophile is D^+ the product is **5.50**, as expected if attack is on the C—C bond (cf. Fig. 5.10). But in basic solution in D_2O the inversion product (**5.51**) is obtained.

The C—Si bond (see above, page 44) has a higher HOMO than C—H, and organosilicon compounds can also undergo electrophilic substitution reactions: as shown by the gradual evolution of methane from solutions of Me₄Si in concentrated sulphuric acid.

5.50   5.49   5.51

Whatever the exact reasons for this reversal of selectivity (and the point is discussed below on pages 74–75) it is clear that for electrophilic attack on cyclopropanes also there is a fine balance between the two possible mechanisms shown in Figs 5.10 and 5.11.

5.6 Suggestions for further reading

Every organic textbook discusses nucleophilic substitution reactions at saturated centres, and every general text on mechanistic organic chemistry devotes a chapter to the reaction. For a good introduction to electrophilic substitution see J. M. Fukuto and F. R. Jensen, *Accts. Chem. Res.*, 1988, **16**, 177–184.

A useful guide to relevant Organosilicon Chemistry is I. Fleming, J. Dunogués and R. Smithers, *Organic Reactions*, 1989, **37**, 57.

There is a vast literature on non-classical carbonium ions, detailing the sometimes fascinating consequences of σ-delocalisation: a useful summary can be found in T. H. Lowry and K. S. Richardson, *Mechanism and Theory in Organic Chemistry*, Harper and Row, New York, 1987.

5.7 Problems

5.1 Explain the relative rates of the reactions with iodide anion (in acetone at 50°) of the alkyl chlorides in the Table below.

n–BuCl	1	MeCl	200
$CH_2=CHCH_2Cl$	79	$PhCH_2Cl$	200
$PhCOCH_2Cl$	100,000	$MeCOCH_2Cl$	36,000
$ClCH_2Cl$	0.2	$MeOCH_2Cl$	920

5.2 Suggest a likely mechanism for the rearrangement in solution of the phosphate triester **C1**.

C1

5.3 The rate constant for the cyclisation of **C2** is a few times smaller than the (second order) rate constant for the reaction of $PhCH_2Cl$ with $PhCH_2NH_2$, and thus thousands of times slower than expected for the formation of a 5-membered ring. Suggest an explanation.

C2   **C3**

5.4 The base catalysed reaction of **C3** gives an ether and a sulphide (a thioether) as products. What are their structures?

5.5 Explain the contrasting results of electrophilic attack (E^+ is a generalised electrophile) on allyl and vinyl silanes (**C4** and **C5**, respectively).

C4

C5

5.6 Explain the contrasting products from the acid-catalysed cyclisation of the closely related epoxides **C6** and **C7**.

C6

C7

5.7 Compound **5.48** (page 46, above) was described as 'carefully designed' to demonstrate inversion of configuration in electrophilic substitution at saturated carbon. Identify and explain the relevant 'design features' in this structure.

6 Additions and eliminations

In this chapter we discuss a large number of important reactions, which have in common the interconversion of sp^2 and sp^3-hybridised centres. In some cases the overall reaction is substitution rather than addition or elimination: substitution at sp^2-carbon generally involves an addition-elimination mechanism, and substitutions at sp^3-centres carrying more than one heteroatom may go by an elimination-addition route.

The more closely we look at the orbital interactions involved the clearer become the underlying similarities between different mechanisms. Particular addition reactions are of course the microscopic reverse of eliminations, and E1 and S_N1 reactions have a first step in common. But stereoelectronic effects on some E2 reactions show marked similarities to stereoelectronic effects on S_N1 reactions, and we have seen in the previous chapter how S_N1 reactions can merge imperceptibly with S_N2 reactions at one end of the mechanistic spectrum.

Addition reactions going by ionic mechanisms involve the addition of both a nucleophile and an electrophile, almost always in separate steps. The usual classification into nucleophilic and electrophilic addition is based on the main HOMO/LUMO interaction in the rate determining step. For additions to ordinary alkenes this involves the interaction of the π-bonding orbital with the LUMO of the electrophile. We look first at reactions of unsaturated systems where the π^*-orbital plays the major role.

6.1 Nucleophilic addition to unsaturated carbon

Perhaps the simplest, and certainly the most familiar addition reaction is the addition of a nucleophile to a carbonyl group.

The orbital interactions involved are discussed in Chapter 2. The HOMO is the lone pair of the nucleophile, the LUMO is $\pi^*_{C=O}$, and the requirement for optimal orbital overlap defines the geometry of approach (Fig. 6.1). When the carbonyl compound has planar symmetry, attack from above and below the plane are equivalent, but in less symmetrical systems 'π-face selectivity' is of considerable synthetic importance.

Addition to C=O

A much-studied example is addition to the C=O group of substituted cyclohexanones. If the conformation is fixed, as in **6.1**, different amounts of axial and equatorial alcohols will be produced. Thus reduction of **6.1** by

Fig. 6.1

borohydride gives a 3:2 preférence for axial attack, to give the equatorial alcohol **6.2e**.

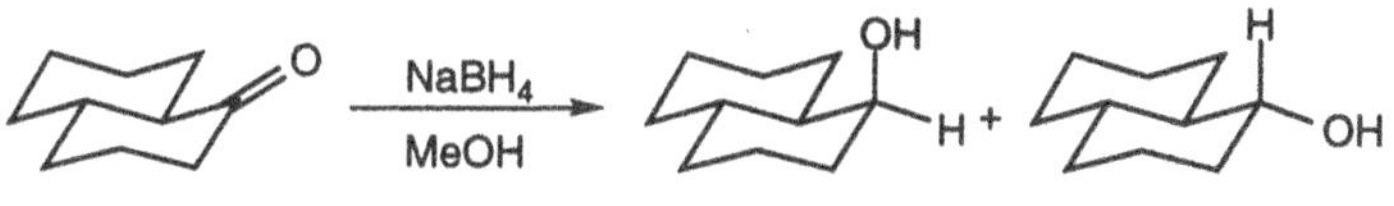

6.1 **6.2a** **6.2e**

The possible involvement of stereoelectronic effects in determining face selectivity has been much discussed in recent years. Different substituents on the carbon atom α to the C=O group can in principle make the two faces electronically different. (To be precise, if the α-carbon is a chiral centre the two faces are diastereotopic.) For example, the developing σ-bond will interact with the C—X bond antiperiplanar to it (**6.3**) on the α-carbon (Fig. 6.2). (Compare the stabilising interactions identified for the S_N2 transition state next to C=O, discussed in the previous chapter, page 40). If the C—X bond is a good σ-acceptor (X strongly electronegative) the interaction of the σ-bonding orbital of the part-formed bond with σ^*_{C-X} will be favourable, and addition will be favoured in this conformation.

π-face selectivity for addition to C=O usually depends primarily on steric effects. The Felkin–Anh model predicts a preference for addition to the face anti to the largest substituent on the α-carbon, as shown below (L, M and S refer to the relative sizes – large, medium and small – of the substituents).

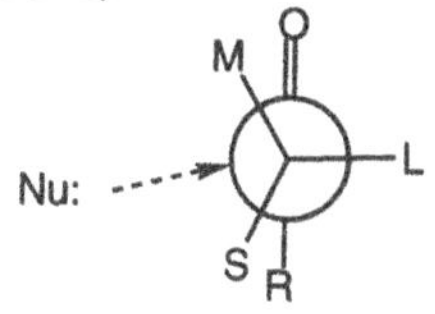

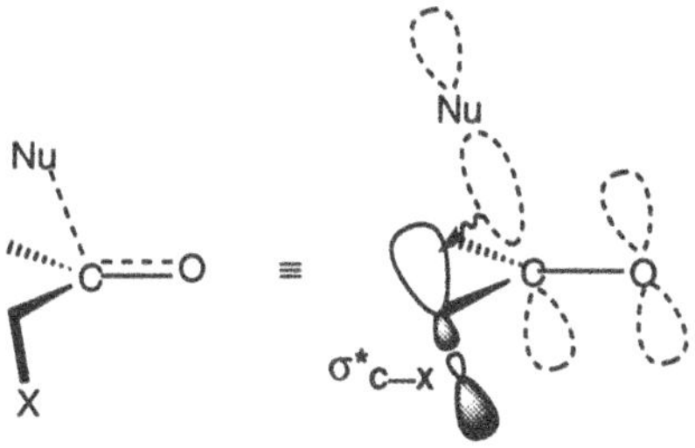

6.3

Fig. 6.2

This sort of analysis can be applied to any system where a nucleophile, an electrophile, or even a radical adds to a π-bond with an adjacent chiral centre. Different ways of thinking about the reaction are possible, but should give the same answer. For example, one could consider the interaction with σ_{C-X} in the ground state, which makes the two faces of $\pi^*_{C=O}$ non-equivalent

This argument immediately raises the question: why should this partial bond not take advantage also of σ-delocalisation in the reverse direction, interacting as an acceptor, with σ_{C-X} as donor? Small stereoelectronic effects in this direction are in fact observed in the reactions of the adamantanone system **6.4**. Substituents Y, which look too far away from the reaction centre to affect it directly, nevertheless affect the stereoselectivity of addition to the carbonyl group. It is found that for electron-withdrawing substituents the main product is **6.5E**, where the hydride nucleophile has come in *syn* to Y. The effect is reversed for electron-donating substituents, confirming that the effect is electronic.

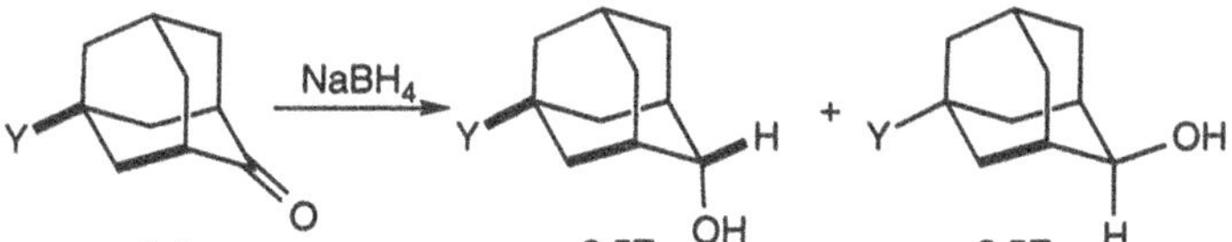

6.4 **6.5Z** **6.5E**

Bonds in antiperiplanar relationships referred to in the text are marked in bold in **6.4** and **6.5**.

One explanation is that the stereoelectronic effects of (two pairs of) C—C bonds α to the C=O group are differentiated by the remote substituent Y. If Y is electron-withdrawing it makes the adjacent C—C bonds – including in particular the bonds antiperiplanar to C—Y) – poorer σ-donors. This would reduce any stabilising σ-delocalisation, making addition preferred antiperiplanar to the other C—C bond, leading to **6.5E**.

Though these effects are significant, they are not large in these ketones, and suggest that the preference for axial addition in cases like **6.1**, without strongly electronegative substituents, is unlikely to be due to the much smaller difference in σ-donor capability between σ_{C-C} and σ_{C-H}. Ordinary steric and conformational effects are generally sufficient explanation. But in systems with stronger electron-demand stereoelectronic effects will be expected to be more significant.

Nucleophilic addition reactions are not limited to C=O groups, and similar stereoelectronic considerations apply to additions to O-protonated (and alkylated) carbonyl groups, C=NR and C=N⁺R₂, and to conjugate additions to αβ-unsaturated derivatives of all of these. Michael addition reactions (Fig. 6.3), for example, involve the interaction of the lone pair of the nucleophile with the LUMO of an αβ-unsaturated carbonyl compound, which is a π*-orbital much like that of C=O, with its largest coefficient at the terminal carbon (Fig. 6.3).

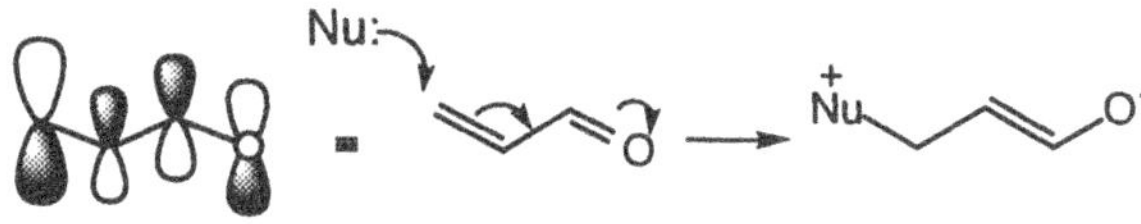

Fig. 6.3

The stereochemistry of nucleophilic addition to π-bonds

The addition of a nucleophile to a carbonyl group generally produces an oxyanion, which is protonated on work-up: or, occasionally, reacts further with an electrophile. In either case the stereochemistry of addition across the C=O bond is not generally defined. But in systems with substituents at both ends of the π-bond the stereochemistry of the reaction can be defined, providing direct evidence about the stereoelectronic effects involved.

A broad generalisation is that there is a preference for the new bonding or non-bonding orbitals formed in the addition reaction to be antiperiplanar, or trans, to each other. This generalisation does not necessarily apply to particular situations or even conditions, but is evidence for a significant degree of stereoelectronic control, and mirrors the preference for antiperiplanar geometry in the reverse, elimination reactions.

The simplest cases involve additions to triple bonds, where the stereochemistry of the double bonds produced is stable and well-defined. Simplest of all are the reactions of the (slightly exotic, very reactive) nitrilium compounds (e.g. **6.6**), where the incoming nucleophile and the lone pair produced end up trans to each other (**6.7**).

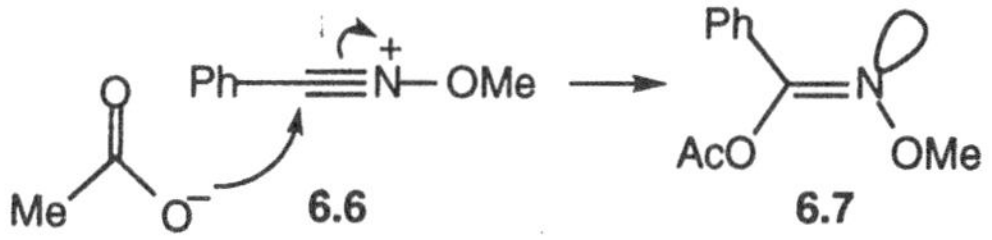

Addition to C≡C behaves similarly, as shown by the product (**6.10**) from the addition of methoxide to phenylacetylene (**6.8**), which also has the original substituents on the triple bond trans to each other in the product; showing that the carbanion (**6.9**) formed in the initial, rate-determining step, again has the lone pair trans to the incoming nucleophile.

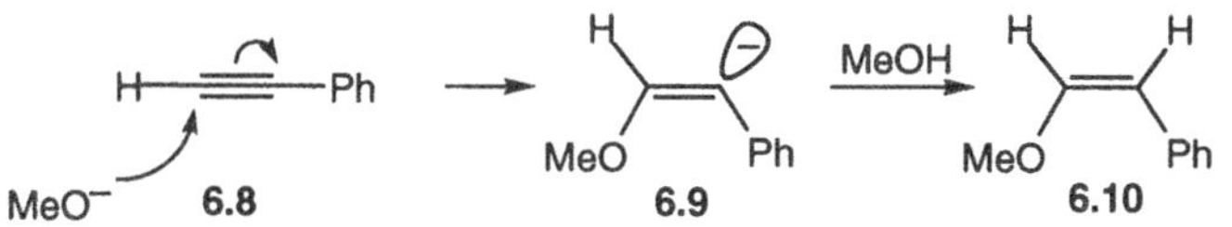

Note that the preferred carbanion **6.9** is stabilised by the favourable antiperiplanar interaction of the localised non-bonding electron-pair with σ^*_{C-OMe}.

Addition to a carbon-carbon double bond results in a conformationally flexible system. Furthermore, the first-formed intermediate is less stable, because a carbanion at sp^3-hybridised carbon is less stable than at sp^2-C, and not likely to form unless delocalised. This makes it more difficult to identify the stereochemistry of the final product with the kinetic product of the first step. Where this is possible trans-addition again seems to be the rule, at least for the addition of hetero-nucleophiles. One of several examples is the base-catalysed addition of lithium thiolates to αβ-unsaturated amides and esters (e.g. **6.11**) in the presence of an excess of the acidic thiol. The product of trans addition is obtained from either isomer (for example, the E-isomer, **6.11**, gives the erythro-product, **6.12**), so the reaction is stereospecific: but this does not necessarily tell us about the stereochemistry of the addition step itself.

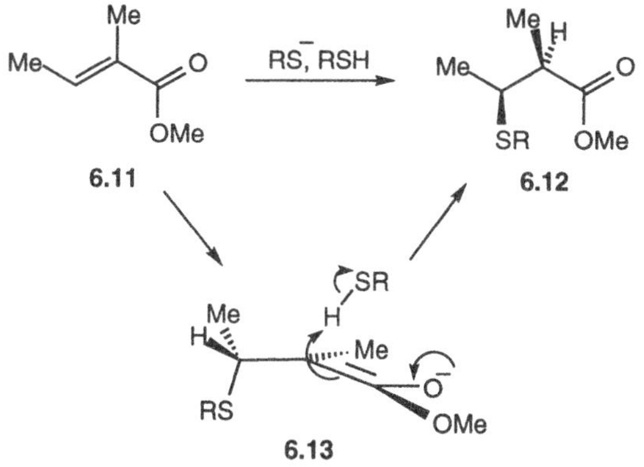

6.13

Fig. 6.4

This result depends on the presence of an excess of the thiol, which accelerates the protonation step and reduces the lifetime of the enolate. Under more basic conditions the reaction is stereoselective only - trans addition is observed, but both E and Z isomers of the ester give the same mixture of products.

It is often a useful simplification to consider a complicated transition state from the point of view of the reverse reaction, in this case the deprotonation of the product **6.12**, followed by the elimination of RS⁻ from a carbanion. Both steps seem intuitively more reasonable this way round; which is discussed in more detail below (page 67).

In fact, under these conditions the stabilised (ester-enolate) carbanion (**6.13**) is a full, though short-lived intermediate, so the stereochemistry of the product is determined not by the preferences of the addition reaction but by the second, protonation step (Fig. 6.4). The situation is very like that described above (Fig. 6.2), where a C—X bond in the corresponding position can affect the π-face selectivity for addition of a nucleophile to the carbonyl group. But here the HOMO/LUMO relationship is reversed, and an electrophile is adding to an electron-rich π-system. So it appears to be favourable to have a σ^*_{C-SR} acceptor orbital antiperiplanar to the electron-rich part-formed bond in this case also (Fig. 6.5).

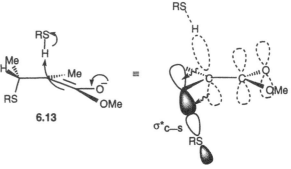

Fig. 6.5

Addition to C=NR₂⁺ and C=OR⁺

N.B. Lone pairs make smaller steric demands than other substituents, and their position generally has to be inferred indirectly, from where the other substituents are. Amine stereochemistry in particular can easily be reversed by the nitrogen inversion mechanism. So discussions of the behaviour of lone pairs in reactions needs extra care.

When a nucleophile adds to an iminium ($C=NR_2^+$) or $C=OR^+$ system the positive charge is neutralised and a lone pair formed. Extrapolating from the results so far in this section suggests that the lone pair might be expected to develop trans, or antiperiplanar, to the incoming nucleophile, and there is a good deal of evidence that there is a stereoelectronic preference to this effect. For example, the borohydride reduction of the bicyclic iminium salt **6.14** gives the trans-fused system **6.15**.

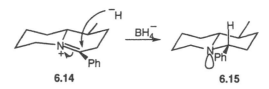

The stereochemistry of this sort of reaction only becomes apparent when the system is conformationally fixed. Delivery of hydride to the equally accessible bottom face of **6.14** would have given a less favourable conformation, and this preference must already be felt in the transition state of the reaction. A convenient way of thinking about the stereochemistry of reactions of this sort is illustrated in Fig. 6.6.

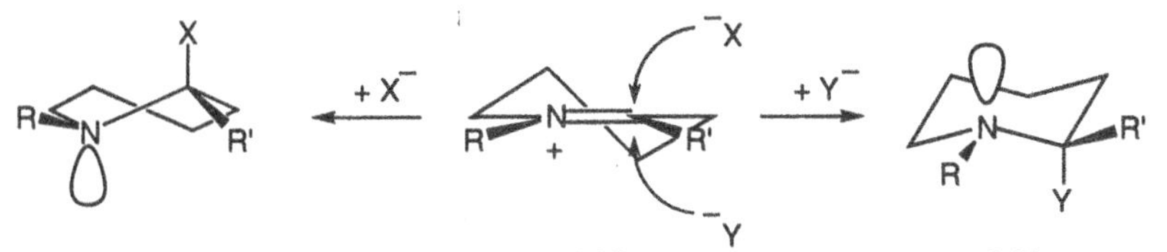

Fig. 6.6

N.B. The argument illustrated by Fig. 6.6 assumes that the conformation of the iminium system is fixed as shown (**6.16**). If it is not, conformational inversion of the ring will make the top and bottom faces equivalent, and the two products will simply be enantiomers.

Addition of the nucleophile (identified as X) to the top face of **6.16** leads to a twist-boat product conformation (**6.17**). This may adapt, for example by inversion of configuration at nitrogen: but initially at least is less favourable than the product of addition of Y to the bottom face. This is formed in the stable chair conformation (albeit with the incoming nucleophile in the axial position). The lone pair is axial, and likely to remain so since the substituent R on the iminium system will prefer to stay equatorial.

Similar considerations apply to addition to C=OR$^+$. This is a high-energy species, and obtained most easily under preparative conditions when stabilised by a second oxygen. Thus alkoxides add to bicyclic cations like **6.19** to give specifically the products (**6.20**) of axial attack. Similarly the reactions of Grignard reagents with cyclic orthoesters **6.21**, which are thought to involve similar dioxacarbenium ions (**6.22**), give the product (**6.23**) with the incoming alkyl group axial.

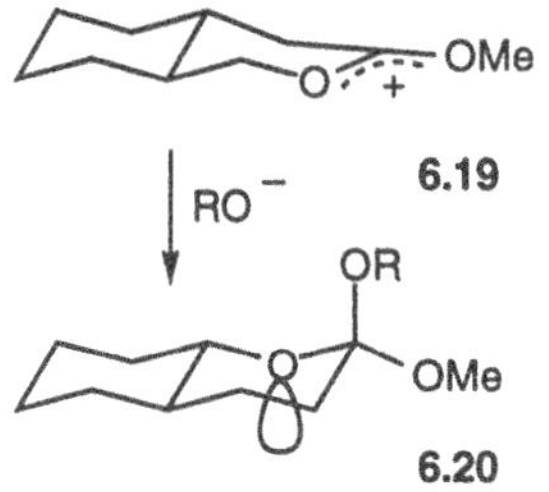

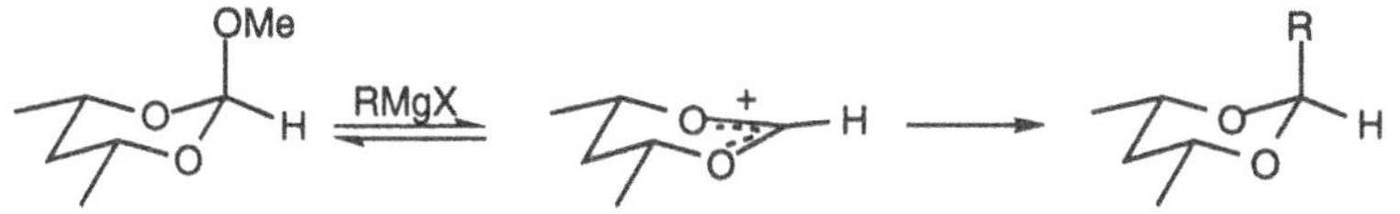

6.2 Electrophilic addition to C=C

The LUMO of a typical simple electrophile has p-symmetry: in the case of a cation or other Lewis acid this is a low-level vacant atomic orbital, while for reagents like proton acids and halogens it is the vacant σ*-orbital of the bond to the electrophilic (proton or halogen) centre. The HOMO is the π-orbital of the double bond, so the basic requirements for efficient HOMO/LUMO overlap favour perpendicular approach of the electrophile (i.e. perpendicular to the plane defined by the bonds of the alkene). The most general picture for a symmetrical alkene is shown in Fig. 6.7:

Fig. 6.7

The exact structure of the initial product (**6.24**) depends on both the alkene and the electrophile, but for the addition of bromine (X = Br), which can form a stable bromonium ion, it is a full intermediate, which can be isolated in favourable cases. In fact for any element X which has lone pair electrons a bridged intermediate is possible. This accounts very simply for the observed trans stereochemistry of addition in such cases: the intermediate **6.24** is opened by the nucleophilic part of the adding reagent, from the back as expected for an S_N2 reaction (Fig. 6.8). This chemistry is familiar, and no new stereoelectronic effects are involved.

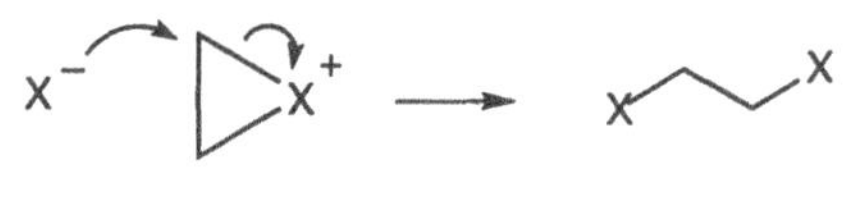

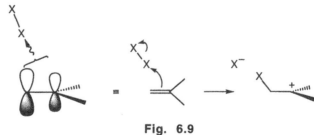

6.24

Fig. 6.8

When a stable bridged structure is not possible, for example for X = H, the symmetrical species **6.24** is no longer an intermediate - i.e. an energy minimum - but a high energy species, probably close in structure to the transition state for addition (or for a 1,2–shift of X - see Chapter 7). The contribution from the rest of the original alkene structure then becomes crucial, and at the other extreme from **6.24** a carbenium ion may be a full intermediate. In this situation the π-system, and the transition state for electrophilic attack, can become very unsymmetrical (Fig. 6.9). The usual curved arrows which indicate the regiochemistry of addition give a good idea of this.

Fig. 6.9

Whether the overall addition reaction is stereospecific therefore depends on the details of the mechanism. We will not go into these details here.

π-Face selectivity

The stereochemistry of electrophilic addition to ordinary alkenes is of special interest for chiral synthesis. The regiochemistry is generally straightforward, controlled predictably by the substituents on the double bond. But π-face selectivity is a more complex problem, especially in acyclic systems. It is subject to the usual array of geometric, steric and stereoelectronic effects, and is a good illustration of the power - and the limitations - of current theory.

We have met the basic problem twice already. It is important in the context of nucleophilic addition to C=O compounds, and arises also in the second step of Michael addition reactions, when the enolate anion first produced reacts with an electrophile. The two faces of the π-system become different (diastereotopic) when there is a chiral centre in the molecule, and subject to stereoelectronic effects when this is in the allylic position, attached directly to the double bond.

When a substituent Y in the allylic position is strongly electron-donating or withdrawing a significant π-σ^*_{C-Y} or π*-σ_{C-Y} interaction is possible if the geometry is right: that is, if the C—Y bond can line up more or less parallel with the adjacent π-orbital (Fig. 6.10).

Fig. 6.10

Whether this conformation is accessible will depend on the other substituents on the double bond. Conformations in this situation are controlled primarily by steric interactions, though any stabilising stereoelectronic interaction will make a contribution to the equilibrium situation.

The most stable conformations about the allylic bond (**6.25** and **6.26** in Fig. 6.11) are staggered with respect to the substituent on C(2), and have the smallest substituent on C(3) eclipsing the C=C double bond. The most significant remaining steric interaction is allylic 1,3-strain, between the bonds marked in bold in Fig. 6.11. The 1,3 interaction between H and a primary alkyl group is not large, and up to 25% of **6.25*** can be present at equilibrium for the monoalkyl alkene. A 1,3 interaction between two alkyl groups is much larger, and **6.26** is the only significant conformation present for a Z-alkene. (The 1,2-eclipsing interaction with the substituent on C(2) is a major disadvantage for conformation **6.26**** even when two hydrogen atoms are involved.)

Conformational energies for rotation about a single bond depend on the torsion angles between the bonds at the two ends, reaching a maximum for each eclipsed conformation. However, the minima are reasonably shallow, and variations of the order of ±30° are tolerated without big effects on the energy of the system.

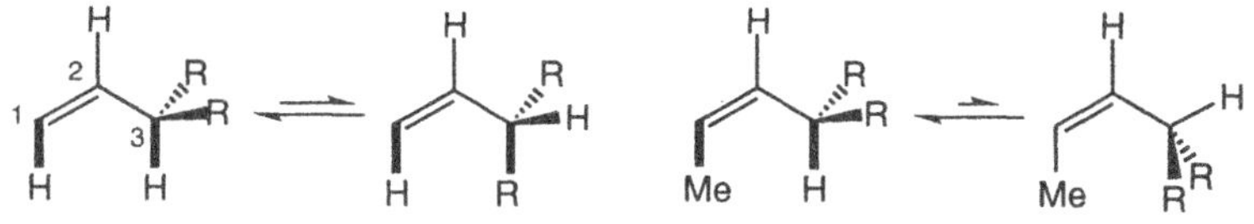

Fig. 6.11

The conclusion in both these cases is that conformations with a C—X bond perpendicular to the plane of the π-system, which require only a 30° rotation from the minimum energy conformations (**6.25** and **6.26**) shown in Fig. 6.11, are quite readily accessible in acyclic systems, in the absence of sterically very demanding substituents.

Electrophilic attack is favoured by electron-donating substituents, and good σ-donors in the allylic position favour attack on the opposite face of the double bond. The clearest cases involve allylsilanes (Y in Fig. 6.10 = SiR_3), where the steric effects of the large trialkylsilyl group hinder the approach of the electrophile and reinforce the stereoelectronic preference. Thus in the alkylation of **6.27** the t-butyl group adds on the opposite side of the double bond from the trimethyl silyl group for both E and Z-isomers, resulting in complete 'transfer of chirality.'

Note that in Fig. 6.12 the Me₃Si group is lost, so the single stereogenic centre in the starting material is replaced by a different single stereogenic centre in the product. For 100% transfer of chirality a pure enantiomer of starting material must be converted to a pure enantiomer of the product.

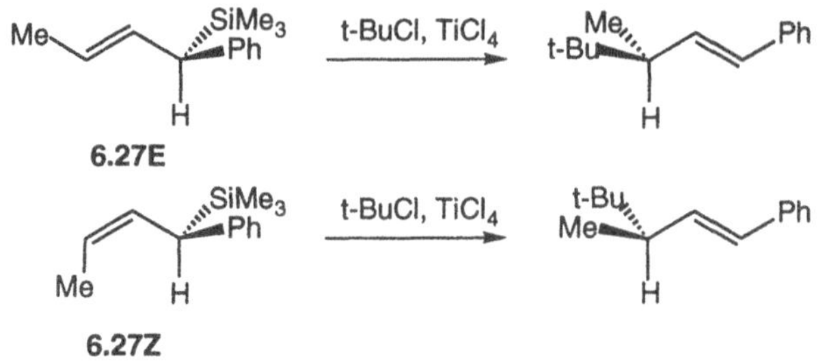

Fig. 6.12

Results are less clear-cut with electron-withdrawing allylic substituents, which should be deactivating, and for which steric and stereoelectronic effects are in opposition. (Strong effects could also affect the regiochemistry of addition.) But there is some evidence for long-range effects in the addition of electrophiles to substituted methylene-adamantanes (**6.28**), where simple steric effects can be ruled out.

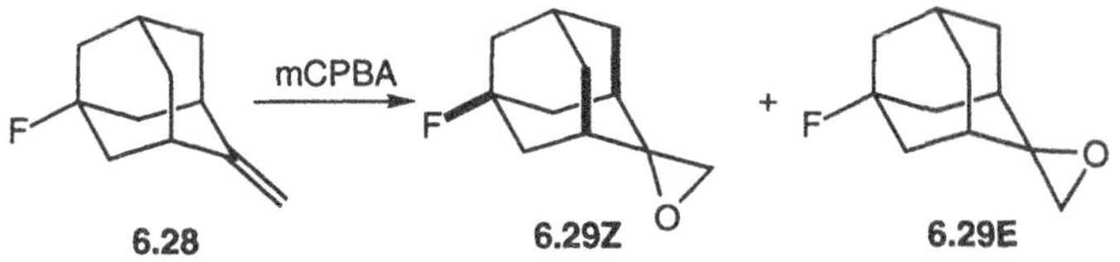

Thus the major product (in a 2:1 mixture of epoxides **6.29**) of epoxidation with *m*-chloroperbenzoic acid was the Z-isomer, in which attack has taken place anti to the two C—C bonds not antiperiplanar to the remote fluorine. This is consistent with a preference for attack anti to the better σ-donor orbitals, in this case the C—C bonds marked in bold.

These results can be rationalised by the orbital interactions shown in Fig. 6.13.

As usual overlap is most efficient between parallel orbitals. The partial bond to the electrophile E⁺ is relatively electron-deficient, so interacts most strongly with σ_{C-Y}. The better this orbital is as a σ-donor, the more the transition state is stabilised, and thus the more the product of attack on the opposite face is favoured.

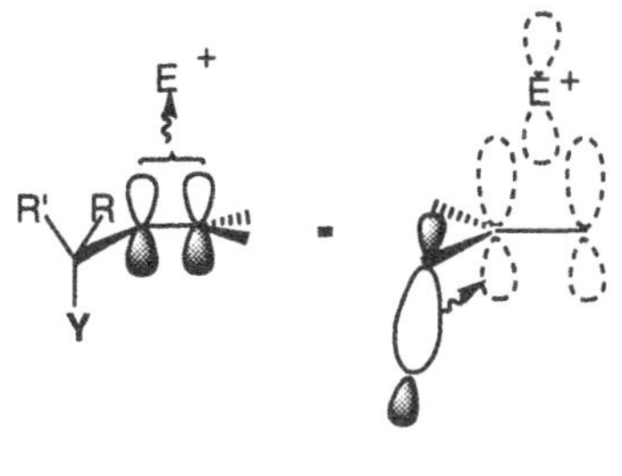

Fig. 6.13

Electrophilic attack on enols and enamines

When there is a powerful π-donor on a double bond, such as NR_2, OR or OH, and especially O^-, the energy of the HOMO is raised significantly, electrophilic attack is particularly easy, and the regiochemistry is unambiguous (Fig. 6.14). π-Donation from the oxygen or nitrogen (also, to a lesser extent, from sulphur) stabilises positive charge on the carbon atom next to it.

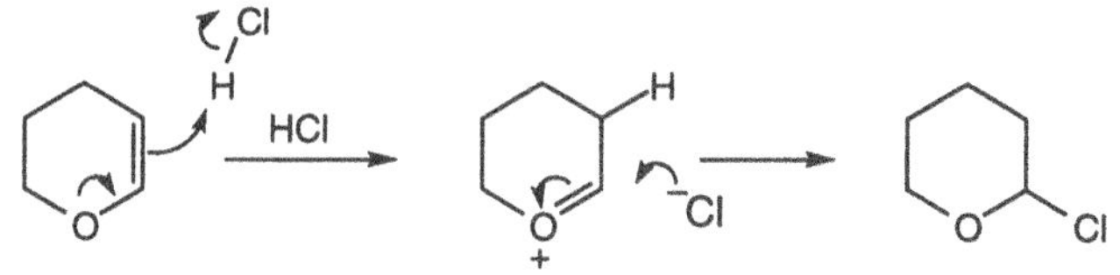

Fig. 6.14

If the geometry restricts or prevents π-donation, reactivity is reduced as expected. Thus the bicyclic compound **6.30**, though nominally an enol ether, is protonated 1400 times more slowly than the parent olefin **6.31**. And although enamines are normally very sensitive to acid, the bicyclic compound **6.32** adds HBr only at very high temperatures, to give a product (**6.33**) in which the proton has added next to nitrogen. In both cases the lone pairs on the potential π-donors are held by the rigid framework in the plane of the π-system, and are unable to stabilise the positive charge developing on the adjacent carbon atom. Protonation, and electrophilic attack in general, is thus subject to stereoelectronic control.

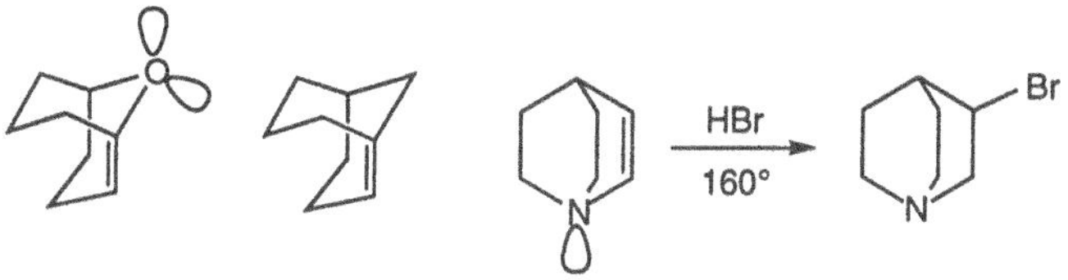

| 6.30 | 6.31 | 6.32 | 6.33 |

In more flexible cases this may be expressed as π-face selectivity, if there remains some conformational restriction. Thus there is a very weak preference for axial protonation of conformationally biased cyclohexanone enolates (7:3 in the case of **6.34**), and an even weaker one for axial alkylation. Axial attack generates the more stable chair form of the cyclohexanone directly, while protonation from the top face produces initially the twist-boat (see above, page 55). Selectivity may increase with increasing rigidity, but any stereoelectronic preference is easily overridden by steric and other effects.

π-Face selectivity induced by the stereoelectronic effects of substituents on the carbon atom next to the one being attacked by the electrophile were discussed in the previous section.

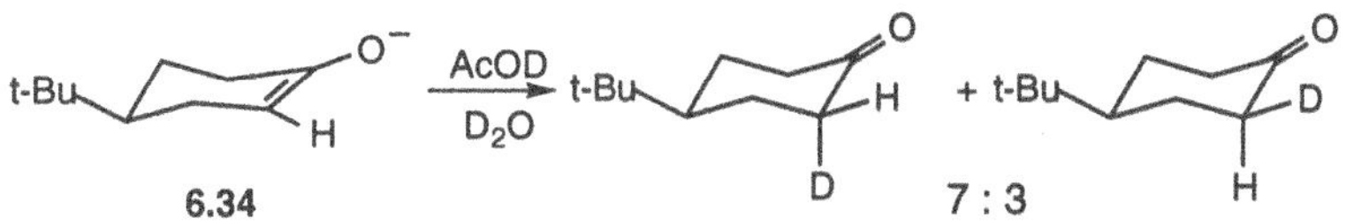

6.34   7 : 3

There is some evidence that this sort of selectivity may be greater for enamines, since alkylation of **6.35** gives more than 90% of axial product (**6.36**) in some cases. One relevant factor may well be reactivity: the faster the reaction, the earlier will be the transition state (an application of the Hammond Postulate: see above, Section 5.4). Earlier transition states for different pathways will resemble the common starting material more closely, so be closer in energy to each other. Selectivity will be reduced as a result, and may be small for this reason; and very small for enolates in particular.

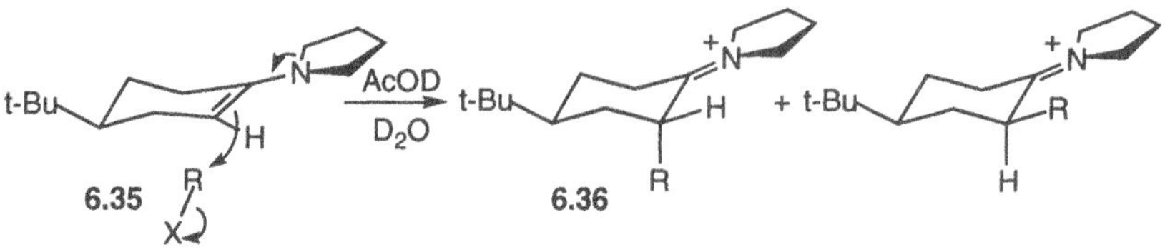

6.3 Intramolecular addition reactions

Intramolecular reactions are subject to the same stereoelectronic effects at the reacting centres as intermolecular reactions. Other things being equal, intramolecular reactions are faster, but this rate advantage disappears for several important classes of reactions. The usual reason is that the preferred geometries of the key orbital interactions are incompatible with the ring being formed. This question of compatibility is the basis of Baldwin's rules (see Chapter 4, page 28), and has already been discussed in Chapter 5 (page 36) in the context of the disfavoured 6-endo-tet reaction.

The most interesting disfavoured addition reactions are 5-endo-trig cyclisations, involving either nucleophilic or electrophilic attack on π-systems to form 5-membered rings. Similar principles are involved in the two cases, and these are discussed first for nucleophilic reactions.

Remember that 5-endo-trig defines the ring size, the sense of the bond being broken, and the geometry of the centre under attack, as explained in Chapter 4.

Intramolecular nucleophilic additions

The example shown in Fig. 6.15 was quoted in Chapter 1: **6.37** cyclises to form the lactam, even though primary amines normally add to the C=C bonds of acrylate esters.

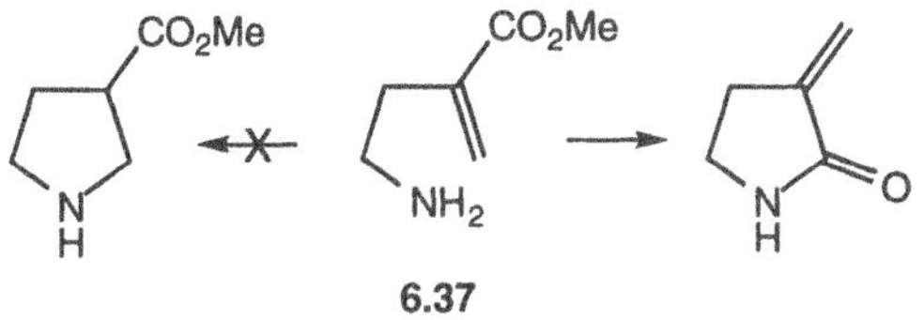

6.37

Fig. 6.15

The selectivity arises directly from the stereoelectronic preferences of the two competing reactions. We know that a nucleophile prefers to attack a π-system from above the plane defined by the bonds to the sp^2-centres (see Figs. 6.1 and 6.3, above). The five atoms forming a 5-membered ring –

including a 5-membered cyclic transition state on the way to one – lie in or close to a common plane. So for a nucleophile to add to the carbon atom of a C=O group five atoms away both must lie in this plane. This means that $\pi^*_{C=O}$ must also lie in the plane if overlap is to be efficient (i.e., if the geometry defined in Fig. 6.1 is to be accommodated in the 5-membered ring). This is possible for the addition of the amino-group to the ester C=O (**6.38**). The reaction is formally 5-exo-trig, since the C=O π-bond being broken is exo to the ring being formed.

6.38

However, this is not possible for the 1,4-conjugate addition of the amino-group to the $\alpha\beta$-unsaturated ester (the reaction would be 5-endo-trig, because the C=C bond broken would be endocyclic to the ring being formed, as shown in **6.39**). The amino-group can approach quite close to the electrophilic CH$_2$, but only in the plane of the π-system (**6.39**). If the amino-group rotates out of plane, to get in position to add to C=C from above, the geometry takes it out of range (**6.40**): and rotation about the C=C double bond is of course prohibited. The result is that this mode of cyclisation is not observed, and the formation of the lactam results, as the fastest reaction that is possible in the circumstances. Many similar cases are known.

6.39   **6.40**

Chemists typically see a rule as a challenge, and look for exceptions to prove it. The 5-endo-trig process does happen, sometimes quite rapidly, and the exceptions make an important contribution to the way we think about stereoelectronic effects. A particularly instructive case is the cyclisation of **6.41**. This compound does not cyclise in base, as is to be expected for a disfavoured 5-endo-trig process, but it does cyclise perfectly well in acid.

6.42   **6.41**   **6.43**

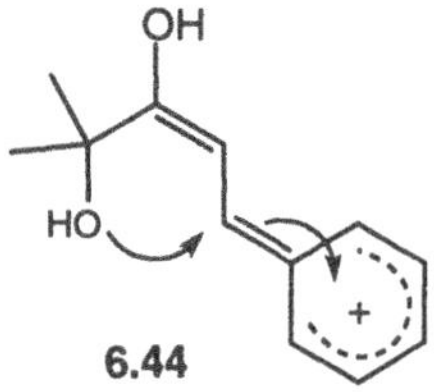

6.44

The original explanation was that the reactive form in acid is the conjugate acid (**6.44**), which has the C=OH⁺ sufficiently stabilised by π-delocalisation into the aromatic ring for the cyclisation to qualify as at least honorary 5-exo-trig. In fact - at least for the 4-methoxy-derivative, which cyclises very rapidly - an alternative mechanism, involving C-protonation, makes this a genuine 5-exo-trig reaction (**6.45**). The effect of C-protonation is to allow free rotation about the original C=C double bond (marked *a* in **6.45**), thus removing the geometrical restriction to overlap.

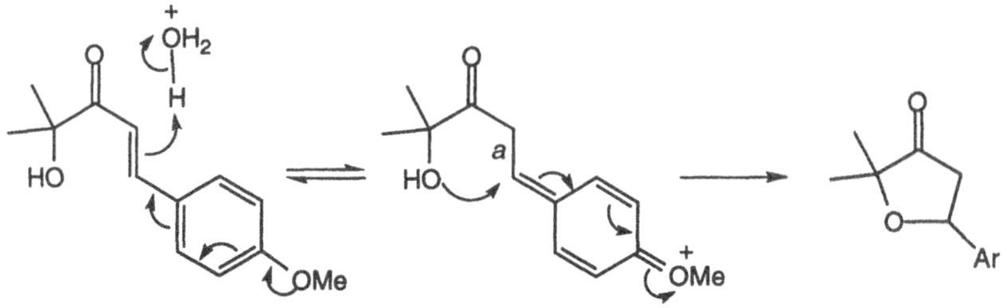

6.45

The ease of the corresponding reactions of alkynes may seem surprising, since the geometry looks distinctly worse for ring closure, and no better for orbital overlap. But **6.46** is readily cyclised even in base.

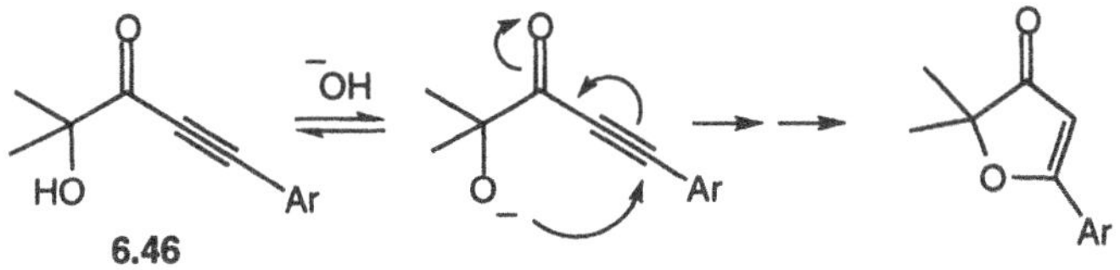

6.46

The important difference here is that an alkyne does have an acceptor, π^*-orbital available in the plane of the 5-membered ring. Even though this is not the one conjugated with the C=O group, it is strongly polarised by it; and alkynes are in any case a good deal more electrophilic than the corresponding alkenes. So cyclisation proceeds, in spite of the potential barriers, because there is a strong enough thermodynamic driving force.

If the driving force is strong enough, 5-endo-trig cyclisations *are* observed. The most familiar involve the formation of cyclic acetals from aldehydes or ketones and 1,2-diols. Here the acid-catalysed cyclisation of the hemiacetal (e.g. **6.47**) must involve a 5-endo-trig process (**6.48**) in many cases. The conclusion must be that even poor orbital overlap can be productive if the reactivity is sufficiently high.

Where there is a choice, 5-membered cyclic acetals may actually be formed faster than 6-membered rings from some sugars: even when the 6-membered ring is thermodynamically more stable.

6.47

In the particular case of formaldehyde acetals it is likely that cyclisation involves the 5-exo-tet displacement of water from the primary centre of **6.49**

6.49

6.48

by way of a transition state stabilised by the adjacent OR group (as discussed in Chapter 5, page 41).

Similar reactions are observed for C=N$^+$ compounds, which are intermediate in reactivity between C=O and C=O$^+$. In fact in one case (**6.50**) a 5-endo-trig cyclisation is faster than a competing 6-exo-trig reaction.

6.50

This is part of a general pattern, in which the stereoelectronic demands of nitrogen seem to be less rigid than those of oxygen in a similar situation. Thus we saw above (page 41) that the C—Cl bond of the chloroether **5.26** is cleaved three times more slowly than that of the corresponding tertiary halide **5.27**; because n_O-σ^*_{C-Cl} overlap is prevented by the geometry of the bicyclic system, and oxygen acts almost exclusively by inductive electron-withdrawal. By contrast the corresponding NMe compound **6.51** is hydrolysed many millions of times faster. Evidently nitrogen, as the better electron-donor, is stereoelectronically more flexible also, and stabilises the developing cation **6.52** significantly despite the unpromising geometry.

5.26   **5.27**   **6.51**   **6.52**

As usual, it is instructive to look at interesting reactions in the reverse direction. For the reverse of the 5-endo-trig process we can see (**6.53**) how the lone pair of a good electron-donor like nitrogen can achieve quite reasonable overlap with σ^*_{C-O}: and when the leaving group is protonated this becomes a viable reaction for oxygen as a donor (**6.54**). But for a typical delocalised carbanion (the enolate **6.55**) the C—O single bond lies in the nodal plane of the π-system, and π-σ^* overlap is negligible. So even when the carbanion can be shown to be formed, for example by exchange of deuterium from the solvent into the CH$_2$ group α to the ketone, this does not lead to ring-opening.

6.53   **6.54**   **6.55**

Intramolecular electrophilic additions

Since electrophilic attack on π-systems also involves approach from above (or below) the plane, the geometrical constraints already outlined for 5-endo-trig cyclisations apply in much the same way. So we are not surprised that the cyclisation of **6.56** gives exclusively the enol ether **6.57** rather than the cyclopentanone **6.58**.

As before, close approach of the centres with the best matched HOMO/LUMO pair brings the enolate CH_2 group and the CH_2Br group together in the plane of the 5-membered ring. As a result the orbitals which must interact to form a new bond are held perpendicular to each other, with σ^*_{C-Br} in the node of the π-system (**6.59**). The enolate oxygen, on the other hand, has a lone pair in the plane, in precisely the right position to overlap with σ^*_{C-Br} (**6.60**).

6.59   **6.60**

6.4 Elimination reactions

Three classes of β-elimination reaction can be distinguished experimentally – concerted E2 and stepwise E1 and E1cb (via the conjugate base). These mechanisms correspond, respectively, to the reverse of concerted addition, and to the reverse of the stepwise additions involving electrophilic and nucleophilic attack on double bonds. In practice elimination and addition reactions are carried out under very different conditions. But the stereoelectronic effects involved are closely related, and E2 mechanisms can be regarded as merging at the two extremes into E1 and E1cb.

The first step of an E1 reaction is the same as for an S_N1 reaction: the loss of a good leaving group to generate a carbocation. This process has already been discussed in Chapter 5 (page 42), and will not be explicitly considered here. Similarly, the first step of an E1cb reaction is the formation of a carbanion, usually delocalised by an adjacent C=O or similar group. We discuss here E2, and the second step of the E1cb reaction, which are stereoelectronically closely related.

The E2 reaction

In the classical E2 reaction a base removes a proton from one centre, and a leaving group departs from an adjacent atom, in processes which are demonstrably coupled. The preferred geometry is well-established. The two bonds which are broken have to be more or less coplanar, and prefer to be antiperiplanar: though synperiplanar eliminations are readily observed if the geometry is sufficiently biased. The sequence shown in Fig. 6.16 is a good illustration (though it should be borne in mind that the Hofmann elimination does favour syn elimination).

Elimination of trimethylamine from the cyclohexyl derivative is almost exclusively antiperiplanar, presumably mostly by way of the inverted chair form, as shown. Two adjacent axial groups are precisely antiperiplanar, and provide the ideal geometry for the anti-E2 process. The rigid cyclobutyl derivative, on the other hand, cannot provide an antiperiplanar hydrogen, but has two in synperiplanar positions. The cyclopentane derivative can pucker to some extent, but most of the 1,2-relationships are synperiplanar for most of the time. The resulting competition between the stereoelectronically less favourable syn process and the stereoelectronically preferred but geometrically restricted anti-elimination gives almost equal amounts of each (Fig. 6.16).

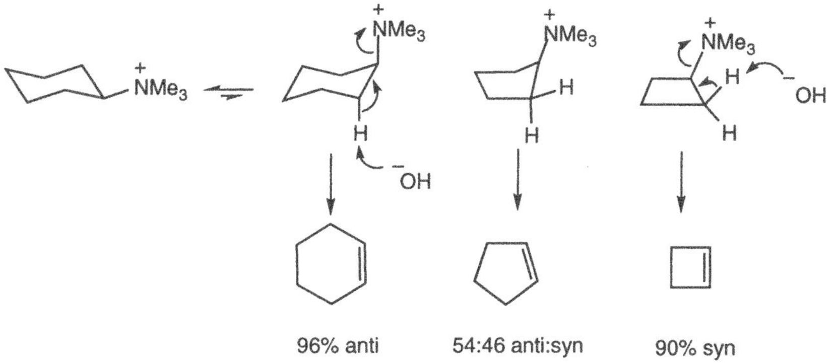

Fig. 6.16

The bond to the leaving group breaks in an elimination reaction because the antibonding orbital becomes populated, by electrons from the σ-donor orbital involved. We have already seen, several times, that overlap between vicinal σ-bonds is most efficient when they are coplanar and parallel, which means antiperiplanar. The picture is intuitively simple - compare the antiperiplanar interaction (**6.61**) with the same interaction in the synperiplanar arrangement (**6.62**) - and supported by evidence of many sorts, from the sizes of NMR coupling constants (Chapter 4, page 25) to the stereoelectronic preferences of reactions.

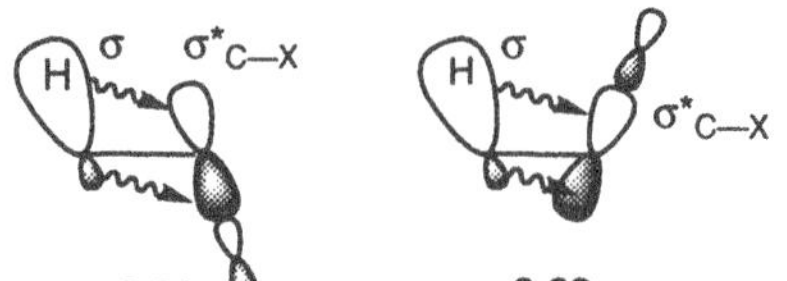

6.61   6.62

The study of structure and reactivity in elimination reactions is complicated not only by the possibility of syn or anti elimination, but also by competition from the S_N2 reaction. For example, restricting the geometry to assess the effect on the E2 reaction is more likely to divert the reaction to other pathways (Fig. 6.17).

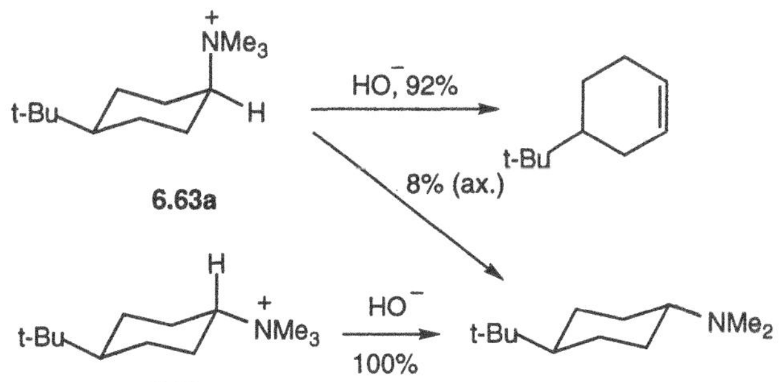

Fig. 6.17

As usual, the situation is simplified where the conformation is fixed, as it is for eliminations across double bonds to give alkynes. Elimination of bromide from **6.65Z**, for example, is over 2000 times faster than the reaction of the E-isomer, which probably adds methoxide first to give, eventually, the dimethyl acetal. Here syn elimination is possible, but must be over 2000 times slower than anti: since the rates of addition of methoxide should not be very different for the two isomers.

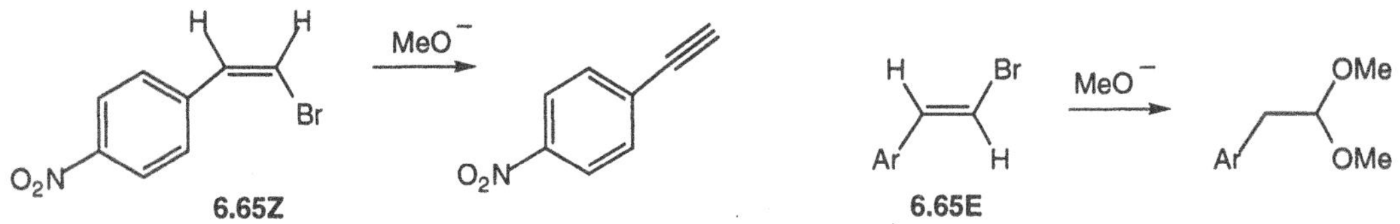

6.65Z   6.65E

It may be that syn elimination is relatively more difficult across a double bond. Overlap in the syn geometry (**6.62**, above) becomes more difficult as the σ-donor and acceptor orbitals diverge further from the preferred parallel orientation. (The angle between the axes of the orbitals concerned, which is zero for the antiperiplanar arrangement and 39° for the syn across two sp^3-hybridised carbons, goes up to 60° for two sp^2-hybridised centres.) A case which may be relevant here is the formation of the high-energy species benzyne (**6.66**) from chlorobenzene, in which the carbanion (**6.67**) is thought to be an intermediate (Fig. 6.18). An sp^3-hybridised carbanion with a vicinal leaving group as good as chloride is not expected to have long enough lifetime to be a full intermediate.

A second relevant factor is the greater thermodynamic stability of a carbanion at an sp^2-hybridised centre.

6.66   6.67

Fig. 6.18

The E1cb reaction

The formation of benzyne in the reaction shown in Fig. 6.18 is noteworthy not just for the unusual nature of product, but as a (slightly unusual) example of the E1cb mechanism. The carbanion **6.67** is the conjugate base of the starting material. For such a species to be a viable intermediate it must have some minimum stability, which usually involves delocalisation of the negative charge, or a poor leaving group, or both. (A stereoelectronic barrier to the loss of the leaving group will also help.)

This means that the two steps, removal of the proton and the elimination of the leaving group, may have quite separate stereoelectronic requirements. The stereoelectronic requirements for the removal of a proton from a carbon acid are simply good overlap of σ_{C-H} with a low energy vacant orbital like $\pi^*_{C=O}$. Those for the elimination step are not very different from those for the E2 reaction: i.e. for optimal overlap the orbitals should lie in the same plane (Fig. 6.19). The geometry of overlap is thus halfway between that for syn and anti E2 eliminations, and if the carbanion is a full intermediate the distinction between syn and anti eliminations may be lost. One reason why it may be retained is negative hyperconjugation (discussed in Chapter 4, page 26): by which orbital overlap of the sort shown in Fig. 6.19 is net stabilising – at least up to the point where bond breaking starts.

This extra stabilisation of the carbanion reinforces the main stereoelectronic effects favouring its formation, and results in an additional preference for removal of the proton antiperiplanar to the leaving group. Thus trans addition products from some Michael reactions undergo H/D exchange under basic conditions preferentially of the proton anti to the original hetero-nucleophile (now the most electronegative group on the vicinal carbon). These reactions go by way of delocalised ester-anion intermediates (**6.69**:

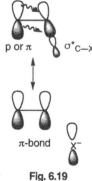

p or π σ^*_{C-X}

π-bond X^-

Fig. 6.19

compare Figs. 6.4 and 6.5, above), in what is the reverse of the first step of the E1cb mechanism.

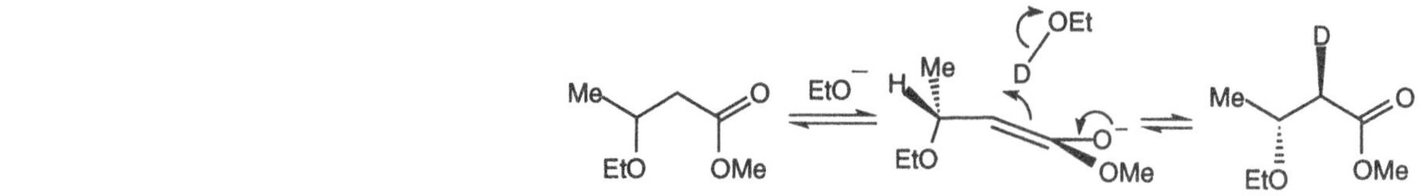

6.68   6.69   6.70

6.5 Substitutions involving addition-elimination and elimination-addition mechanisms

Sequential addition-elimination mechanisms account for almost all substitutions at unsaturated centres. These include aromatic substitution, both nucleophilic and electrophilic, and the whole range of acyl transfer reactions – from the hydrolysis and formation of esters and amides to some Friedel-Crafts reactions. The great majority of these reactions are readily understood on the basis of the principles discussed for nucleophilic and electrophilic additions in the first two sections of this chapter. In terms of stereoelectronic effects the second step is simply the reverse of the first, though in both cases the effects may be modified by particular local conditions.

Addition-elimination reactions

To take nucleophilic aromatic substitution (S_NAr) as a specific example: the initial addition step involves the bonding interaction of the HOMO of the nucleophile with the lowest vacant π^*-orbital of the aromatic system as the LUMO. We do not need to look up the detailed properties of this orbital: we know that it has π-symmetry and a high coefficient at the centre where nucleophilic attack is observed. So the process is comparable to the addition of a nucleophile to C=O, and we are rather confident that it will take place from above the plane of the ring, almost certainly from an angle similar to the Bürgi-Dunitz angle, and fairly close to the symmetry plane through C(1) and C(4) (Fig. 6.20).

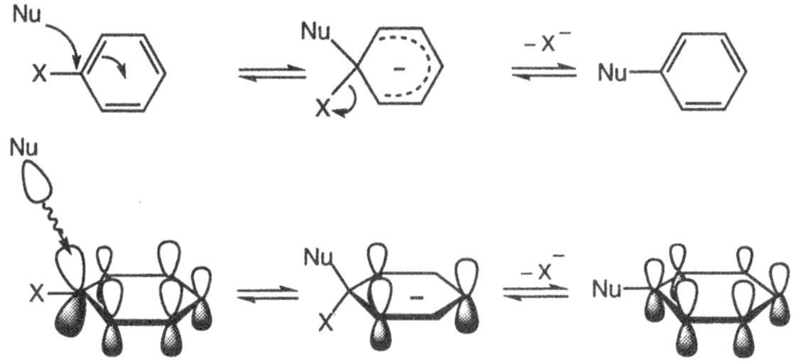

6.71   6.72

Fig. 6.20

Of special interest for reactivity is the structure of the tetrahedral addition intermediate **6.72**. The delocalised π-system is the equivalent of a pentadienyl anion, so the symmetry of the HOMO is known: it has nodes at carbons 2 and 4 (of the pentadiene), and the electron-density is concentrated at carbons 1, 3 and 5 – the positions ortho and para to the site of substitution, as shown in **6.72**. This helps us to understand why nucleophilic aromatic substitution depends on the presence of π-acceptors, like nitro-groups, in one or more of these positions.

The nucleophile and the leaving group X are often both hetero-atoms, so the tetrahedral intermediate may be stabilised further by the effects, including the generalised anomeric effect, which operate on systems with two hetero-atoms on the same carbon atom (see above, page 18). Finally, X is acting as a leaving group, so σ^*_{C-X} is by definition a low-energy antibonding orbital. If Nu has available lone pair electrons these will also assist in the bond-breaking process, as discussed in Section 5.3 (page 41).

It is instructive to try this sort of analysis on other addition-elimination reactions. Reactions at C=O centres are more straightforward, as the negative charge displaced is simply parked temporarily on oxygen (**6.73**): the double-headed arrow in **6.74** is a convenient shorthand for what is going on.

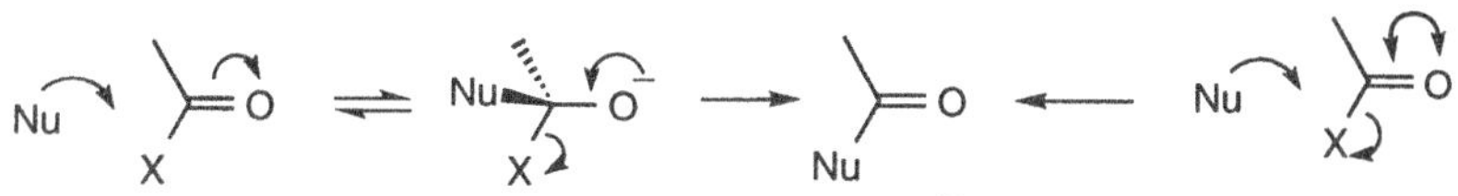

6.73 X^- **6.74**

The fact that substitutions at sp^2-centres go by addition-elimination mechanisms tells us that the LUMO is the π^*-orbital and not the σ^* orbital of the C—X bond being broken, though in principle a concerted S_N2-type of displacement reaction is not impossible. In fact under conditions which might favour this process – very electronegative X, and thus very low-lying σ^*_{C-X} – the mechanism can also change to elimination-addition. Thus the formation of the acylium ion intermediate (**6.75**) (Fig. 6.21) involves the interaction of the (presumably) antiperiplanar sp^2-hybridised lone pair on the carbonyl oxygen with σ^*_{C-X}:

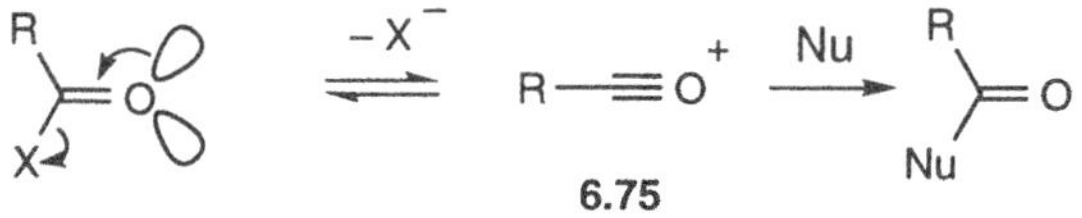

6.75

Fig. 6.21

The stereochemistry of this process is straightforward. For related elimination-addition reactions involving sp^3-hybridised oxygen there are interesting complications.

Elimination-addition reactions

Elimination-addition reactions involving sp^2-carbon have well-defined stereochemistries and it is generally a simple matter to identify the stereoelectronic effects involved. Thus the Z-vinyl chloride **6.76** is converted to the Z-enol ether **6.77** by way of the acetylenic sulphone. Evidently both the elimination and the addition steps involve preferred antiperiplanar orbital interactions.

The stereochemistry is less clearly defined at sp^3 centres, especially where lone pairs are the donor orbitals in the elimination step, and are regenerated in the addition step of the reaction. This is the case for most reactions at acetal centres, and for reactions of related compounds like orthoesters.

Acetal hydrolysis is the most familiar of these reactions. The C—O cleavage step generates a carbocation stabilised by π-type delocalisation of lone-pair electrons from the remaining oxygen, exactly as in the first step of the S$_N$1 reaction (**6.78**, X = ROH$^+$: compare **5.24**, page 41).

We have seen how the reactant prefers the conformation with the lone pair antiperiplanar to the C—X bond, which allows the strongest n$_O$—σ* interaction. The product requires the corresponding conformation **6.79** to allow π-delocalisation. This suggests that an antiperiplanar lone pair is needed for easy C—X cleavage, and that one is also generated when a nucleophile adds to a C=OR$^+$ double bond in the reverse reaction. We have already seen some evidence for these propositions (page 55). Since they apply to both steps of the elimination-addition mechanism it is important to be able to assess the extent of this stereoelectronic control of these reactions.

Important early evidence came from oxidation reactions of orthoformate esters and acetals, in which electrophilic attack by ozone is thought to remove hydride to generate initially a hydrotrioxide (e.g. **6.80**). Compounds like β-glycosides (**6.81**) and the rigid bicyclic acetal **6.82** react, whereas conformationally rigid α-glucosides (based on **6.83**) and acetal **6.84** do not.

The principle of stereoelectronic control in this sort of system was put forward in its simplest form by Deslongchamps, and is discussed in detail in his landmark book (see the Preface to this Primer).

The reactive systems have two lone pairs in positions where they can be antiperiplanar to the C—H bond being attacked, and the similar behaviour of orthoesters in substitution reactions (see above, page 55) gives strong support to the suggestion that they play a key role in the transition state:

specifically, that two antiperiplanar lone pairs are required for easy cleavage of a bond to the central carbon.

Acetal hydrolysis (Fig. 6.22) involves breaking the bond to one of the oxygens, so only one is available as a donor.

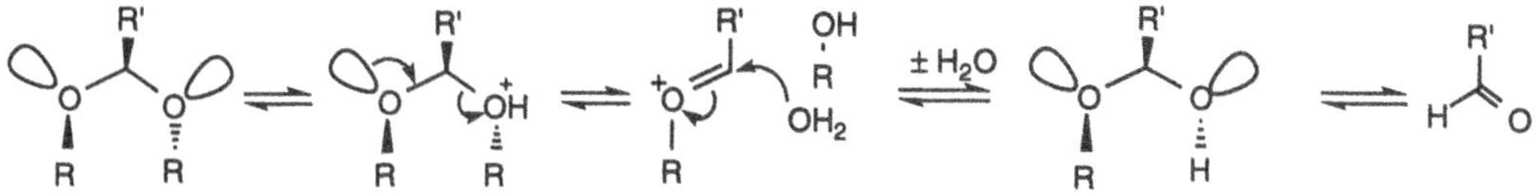

Fig. 6.22

The reaction (**6.86**) should be even more dependent on the efficiency of electron-donation from this single donor, as indeed it is. But activation energies are now higher, and this requirement only affects the observed reactivity to any significant extent when conformations are rather firmly fixed. The theory predicts, for example, that a glycoside or tetrahydropyranyl acetal should be hydrolysed more rapidly when the leaving group is in the axial position (**6.83**, above, rather than **6.81**), because there is a lone pair on the ring oxygen antiperiplanar to the bond being broken.

This is not observed: axial glycosides and tetrahydropyranyl acetals like **6.88** are hydrolysed at about the same rate as their equatorial isomers, and often slightly more slowly. A simple example is the spontaneous cleavage of the oxadecalin acetals **6.88a** and **6.88e**. Here the equatorial isomer **6.88e** is hydrolysed three times faster; no doubt because the conformation is flexible away from the trans bridge. Reaction can occur through a twist-boat conformation **6.88b**, which is readily accessible from **6.88e**, and particularly stable, because it can now take advantage of stabilisation from the anomeric effect. The situation is the one described in Chapter 4 (case 3, page 31). Since the reactive conformation **6.88b** can be attained at little or no cost in conformational energy, and the barrier to the conformational change is small compared with the activation energy for the reaction, the stereoelectronically incorrect isomer can still react easily. (In fact it reacts faster, because the axial isomer is stabilised by the anomeric effect, and the relative rates are determined by the relative ground state energies.)

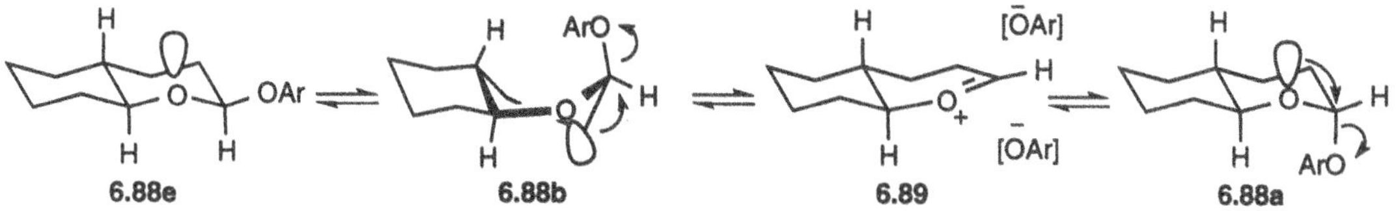

Remember that in the chair conformation of a saturated 6-membered ring, a ring bond. is always antiperiplanar to any equatorial bond.

The stabilised carbocation intermediate (**6.89**) is rapidly hydrolysed to products. The same intermediate is formed from both isomers, differing (while the ion pair stays together) only in the face on which the anion is released.

When the conformation is fixed the magnitude of the contribution from $n_O–\sigma^*_{C–O}$ overlap does become apparent. The bicyclic acetal **6.90**, which is rather rigidly fixed in the equatorial conformation, is extremely unreactive. In contrast, the corresponding axial analogue (**6.91**) is too reactive to prepare, but extrapolations give an estimated difference in reactivity of 10^{13}. As we have seen before in this sort of bicyclic system, overlap between the lone pairs on the ring oxygen and the C—OAr bond in **6.90** – or the vacant p-orbital of any carbocation **6.92** – is prohibited.

6.90	**6.92**	**6.91**

This is an extreme case, where the high energy of the transition state for the cleavage reaction of **6.90** is determined mainly by the high energy of the intermediate **6.92**, and the axial analogue reacts by way of a different intermediate. Where axial and equatorial isomers react by way of the same intermediate reactivity differences are smaller, but can still be substantial; as in the case of **6.93a**, which is hydrolysed in acid 300 times faster than its equatorial isomer **6.93e**.

6.93e		**6.93a**

The conclusion is familiar: a reaction which is thermodynamically favourable will go by the lowest energy pathway available. Stereoelectronic (and other kinetic) barriers affect reactivity only if no lower-energy pathway to products is available.

6.6 Additions to C—C single bonds

Additions to carbon-carbon σ-bonds, in the sense:

$$C-C \;+\; X-Y \quad \rightarrow \quad X-C \;+\; C-Y$$

are rare. The energy balance is not favourable for a reaction in which two σ-bonds are converted simply into two new σ-bonds, unless the C—C bond is particularly weak. And unless the C—C bond concerned is in a ring the reaction is in any case not formally an addition. So it is not surprising that such reactions are limited almost exclusively to cyclopropanes.

Both nucleophilic and electrophilic additions are known, and both resemble the reactions of alkenes in many respects. Nucleophilic addition requires delocalisation of the developing negative charge, just as in the Michael reaction, and the same π-acceptor groups work in both cases. Thus good nucleophiles open esters (**6.94**) of cyclopropane-1,1-dicarboxylic acid to give addition products (**6.95**):

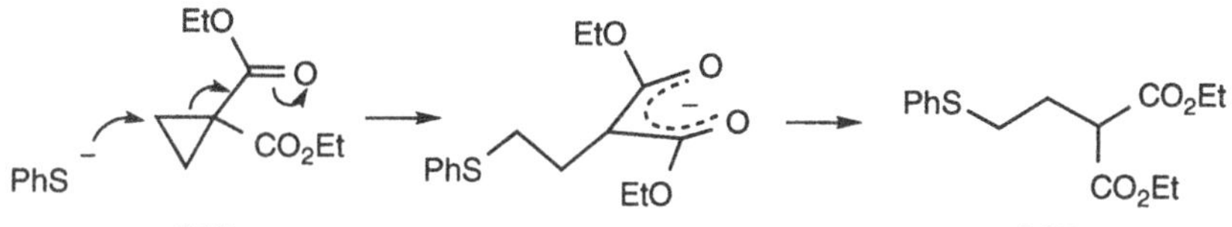

6.94 → → **6.95**

This looks very much like an S_N2 reaction at the carbon atom attacked by the nucleophile, and the reaction does indeed go with inversion at this centre (Fig. 6.23).

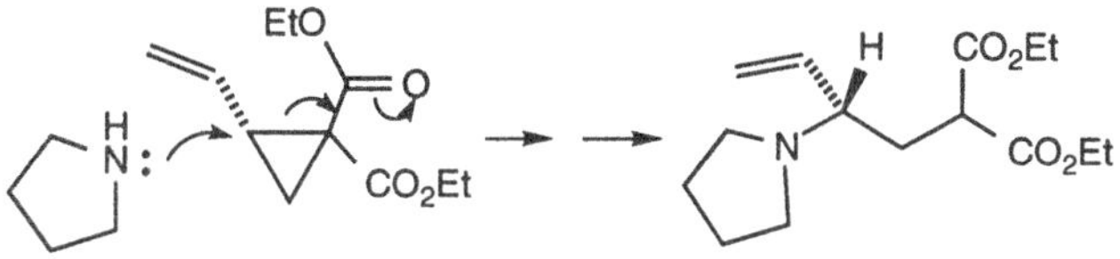

Fig. 6.23

Efficient delocalisation of the pair of electrons displaced (the C—C σ-bond which breaks) is essential. As we saw in Chapter 5 (page 44) this happens most efficiently in the bisected conformation (**6.96**). This stereoelectronic preference explains why the spirocyclic diester **6.97**, which has both C=O groups held by the ring in the desired conformation, is particularly reactive, and thus the ester of choice for synthetic applications of this (1,5-addition) reaction.

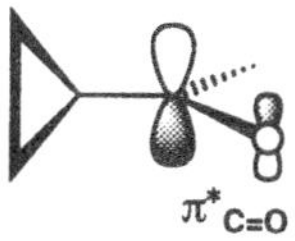

6.96

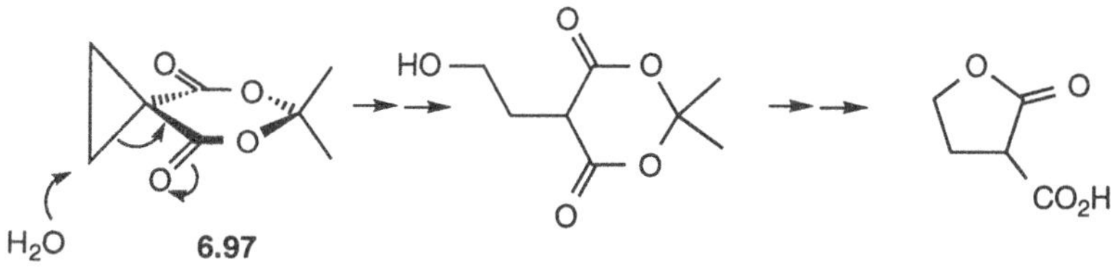

6.97

Typical electrophilic reagents like bromine and the halogen acids add to cyclopropanes, though not so easily as to the corresponding alkenes. As for nucleophilic addition, the two reactions have a great deal in common, but the stereochemistry is quite different. The alternative products from alkenes are the trans or cis adducts, but for cyclopropanes the choice is between inversion or retention of configuration at the two reacting centres. (The regiochemistry is also more complicated because all three bonds of the cyclopropane may be attacked.)

Halogenation generally involves attack by the electrophile on the bent, electron-rich σ-bonds of the three-membered ring. The reaction is very much like attack on a π-bonding orbital as discussed earlier in this chapter (page 55), and involves the interaction of the σ-bond with the σ*-antibonding orbital of the electrophile, as shown for a symmetrical system in **6.98**. (An intramolecular version of this type of interaction involving a vacant p-orbital as the acceptor was discussed in Chapter 5 (page 45)). The C—C bond can then break spontaneously (to form the more stable carbocation); or with assistance from the nucleophilic part of the reagent, leading to retention of configuration at the first centre and inversion at the second: corresponding – for related reasons – to trans-addition to an alkene.

6.98

Other electrophiles, however, may attack at the corner, rather than the edge of the cyclopropane, and here the story becomes more complicated. The result can range from complete retention to complete inversion of configuration. A fascinating case in point is the opening of the cyclopropanol **6.99** derived from tricyclene. Cyclopropanols, like enols, are particularly reactive towards electrophiles, and are opened under mild conditions even in water, to give ketones or aldehydes. The reaction is catalysed by acid or by base, and running the reaction of **6.99** in deuterated acidic solvent showed that protonation leads to the product (**6.100**) with retention of configuration. This is consistent with electrophilic attack on the σ-bond, as shown (and as already discussed on pages 47-48): presumably the interaction with the electrophile is a very unsymmetrical version of **6.98**.

However, when the reaction (now of the more reactive anion) is run in strong base, inversion results (**6.102**). A simple explanation is that the normally weak n_O-σ^*_{C-C} interaction, between the antibonding orbital of the C—C bond of the cyclopropane and the lone pairs of the alkoxide anion, weakens the bond in such a way that the tail of the bonding orbital becomes more prominent. This redistribution of electron-density results in the preferred HOMO/LUMO interaction shown (**6.101**), when the LUMO is σ^*_{O-D} of a weak acid (t-BuOH).

6.7 Suggestions for further reading

Addition and elimination reactions are covered in every organic textbook. A useful mechanistic treatment that complements the discussion in this chapter is Chapter 7 of T. H. Lowry and K. S. Richardson, *Mechanism and Theory in Organic Chemistry*, Harper and Row, New York, 1987.

For relevant Organosilicon Chemistry see I. Fleming, J. Dunogués and R. Smithers, *Organic Reactions*, 1989, **37**, 57, and S. E. Thomas, in *The Roles of Boron and Silicon*, Oxford Chemistry Primer #1, 1991.

Michael addition reactions are reviewed in detail by: P. Perlmutter, *Conjugate Addition Reactions in Organic Synthesis*, Pergamon, Oxford, 1992. R. W. Hoffmann, *Allylic 1,3-Strain as a Controlling Factor in Stereoselective Transformations. Chem. Rev.*, 1989, **89**, 1841 gives a useful discussion of this aspect.

Stereoelectronic aspects are dealt with specifically by K. Mikami and M. Shimizu, *Stereoelectronic Rules in Addition Reactions* in *Advances in Detailed Reaction Mechanisms*, JAI Press, Vol. 3, 1994. For a recent discussion of Baldwin's rules see C. D. Johnson, *Accts. Chem. Res.*, 1993, **26**, 476-482.

6.8 Problems

6.1 Explain how the addition of thiolate to the E and Z-isomers of **6.11** (Fig. 6.4, page 53) can give the same mixture of products under basic conditions.

6.2 Explain the stereoelectronic factors involved in (a) the addition of phenol to the alkyne **D1**; and (b) the formation of **D2** from Z-dichloroethylene and PhS$^-$ (note that E-dichloroethylene does not react under the conditions).

6.3 Discuss how stereoelectronic effects control the regio- and stereochemistry of the reactions of **D3** and **D4** shown below.

6.4 Account for these contrasting results.

7   Rearrangements and fragmentations

So far we have considered almost exclusively short-range stereoelectronic effects, through space or through just one bond. This chapter brings together two groups of reactions of synthetic importance and mechanistic interest, where the key interactions are through two or three bonds.

Ionic reactions happen when a powerful enough electron-donor interacts with a powerful enough electron-acceptor. When two such orbitals are present in the same molecule, *but not in a position to interact directly*, they may still interact strongly enough to cause a reaction. They can do this by involving orbitals of the intervening framework.

We start with a simple example - the interaction of an alkoxide anion with an alkyl halide. Interaction through space leads to an S_N2 reaction. If this happens when both are on the same molecule a cyclic ether is the likely product, from a favoured exo-tet reaction (see page 28). But in some systems, particularly where the geometry prevents the direct interaction, quite different chemistry results (Fig. 7.1).

Rearrangement products

Fragmentation products

Fig. 7.1

Rearrangement (Reaction 2) may be observed even in some conformationally mobile systems, where it must compete with epoxide formation. The corresponding cyclisation (to form a 4-membered cyclic ether) in competition with Reaction 3 in a mobile system is a 4-exo-tet reaction – favoured but particularly slow.

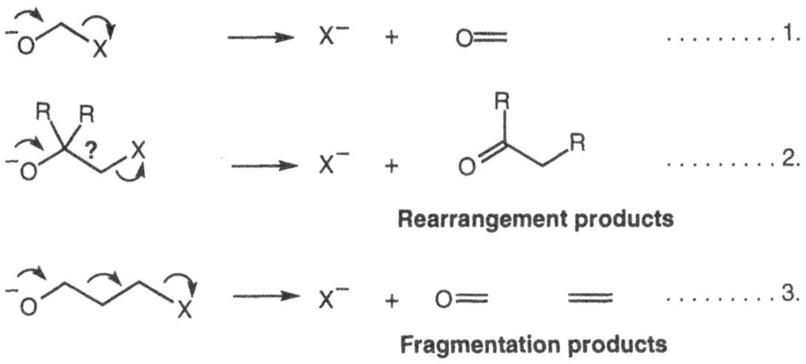

With alkoxide oxygen and halogen on the same carbon the halide ion is rapidly eliminated, to give the C=O group (Reaction 1). If they are on vicinal carbons the formation of C=O and the elimination of halide are still thermodynamically desirable, but no direct HOMO/LUMO interaction is available to bring this about. The reaction requires a specific *rearrangement* of the carbon skeleton (Reaction 2, R = H, alkyl or aryl), as discussed below. Finally, if the two are separated by three atoms (all carbon in the example shown as Reaction 3), breaking the central C—C bond allows the desired formation of C=O and the elimination of halide, in a sort of double elimination. This is a *fragmentation* reaction. Rearrangements and fragmentations provide subtle ways of reorganising the carbon skeleton, predictable since both are subject to well-defined stereoelectronic effects. The synthetic chemist needs to understand them well enough be able to use them creatively - and should never be taken by surprise by either!

To analyse the stereoelectronic interactions involved it is best to start at the leaving group end. The initial interaction is the familiar σ_{C-C}-σ^*_{C-X}, between the best donor σ-bond (of up to nine available on vicinal carbons) and the antibonding orbital of the C—X bond. (Or the vacant p-orbital of a carbocation if the leaving group has already left.) We know that this interaction is strongest in the antiperiplanar conformation (**7.1**), as we saw for example for the E2 reaction (Chapter 6, page 65).

Most rearrangements and fragmentations are electrophilic, driven primarily by the departure of the leaving group. But - as usual - what is important is the difference in energy between the HOMO and LUMO involved: so reaction may still occur with a relatively poor leaving group if the electron-density at the donor end is high enough.

7.1   **7.2**   **7.3**

As the bond to the leaving group breaks (**7.2**) σ^*_{C-X} becomes more electrophilic, until it eventually becomes the vacant p-orbital of a carbocation, stabilised by σ-donation from all available bonds on vicinal carbons. What happens next depends on how effectively positive charge can be accommodated at each of the three carbon centres directly involved - labelled α, β and γ in **7.2**. If the carbocation formed at C_α (**7.3**) is sufficiently stable this first step may define the overall reaction, for example as S_N1 or E1.

If C_γ gives the most stable cation, the C—C bond can break, to generate an alkene (**7.4**). The carbocation formed then reacts further to give products, which depend in the usual way on the conditions. The stereoelectronic effects involved in the stabilisation of this cation, and favouring fragmentation, are discussed in the next section.

Carbocations are stabilised strongly by electron-donating substituents (σ or π-donors). A strong π-donor like the alkoxide oxygen in Reactions 1-3 of Fig. 7.1 will control the overall reaction.

7.2   **7.4** X^-

The most complicated case arises when the carbon centre best able to accommodate the positive charge is C_β, vicinal to the C—X bond. The result in this case is a rearrangement, involving the migration of C_γ from C_β to C_α *with the electrons of the σ-bond* (**7.5** $\rightarrow$ **7.6**). Rearrangement reactions are discussed in section 7.2, below.

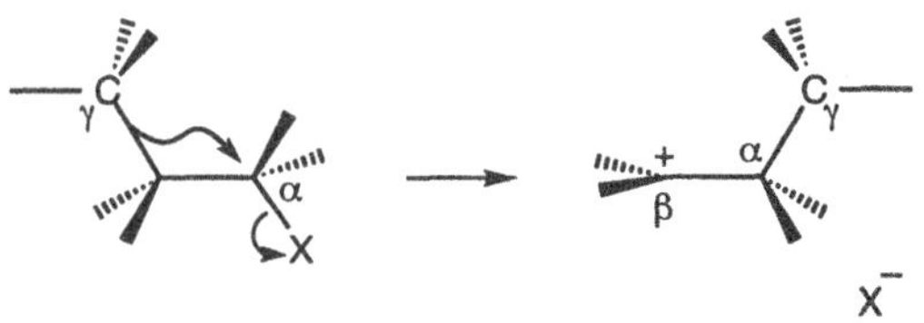

7.5 **7.6**

7.1 Fragmentations

Fragmentation reactions are observed when a strong electron donor interacts with a good leaving group three carbons away. The simplest case is the acid-catalysed fragmentation of a 1,3-diol to an alkene and a ketone or aldehyde (Fig. 7.2). In the reverse direction this reaction is useful in synthesis, when the diol is thermodynamically favoured.

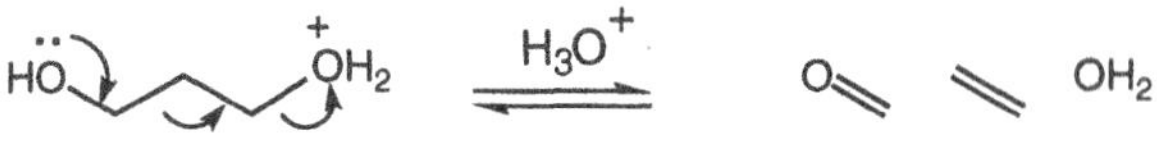

Fig. 7.2

Control is better if one of the oxygens is functionalised, and the reaction provides a useful route to strained medium-sized rings. An example is the fragmentation of the decalin **7.7**, which is converted in base to the cyclodecenone **7.8**.

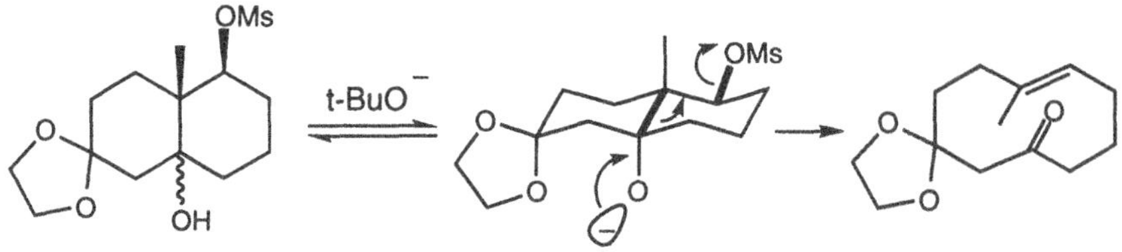

7.7 **7.8**

This is the model for a large number of related reactions, which have in common the interaction of a donor oxygen and an acceptor $C-O$ σ^* or π^*-orbital. They include such familiar cases as the retro-aldol reaction, the decarboxylation of β-ketoacids and the alkaline cleavage of 1,3-diketones. The addition of a nucleophile to a carbonyl group is a simple way to generate a donor oxyanion, and where the addition is sufficiently selective clean reactions are observed. Thus **7.9** gives acetate and cycloheptanone in almost quantitative yield. (The corresponding cyclohexanone derivative gives mainly ring-opening product.)

Note that **7.9** is a 1,3-diketone, and converted in base mainly to the delocalised anion, which is protected from the cleavage reaction.

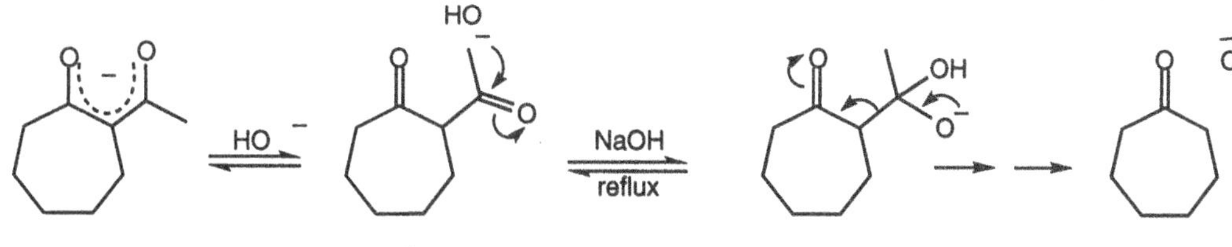

7.9　　　　　　　　　　**7.10**

In all these cases the reaction depends on the geometry of the system. The direction of the lone pairs on the donor O⁻ is not fixed, so the decisive requirement is for efficient overlap between the bond being broken and the terminal acceptor orbital. The requirements for efficient overlap are familiar from previous chapters: the bond being broken should be perpendicular to the plane of the carbonyl group for overlap with $\pi^*_{C=O}$, or antiperiplanar to σ^*_{C-OX}, exactly as for an elimination reaction. Thus the equatorial mesylate **7.7** fragments as shown above (the relevant antiperiplanar bonds are in bold). The axial isomer would react quite differently.

The geometry of the interaction which stabilises the developing cation at C_γ is well-defined for cyclic ethers and amines. Reaction requires good, preferably antiperiplanar overlap for the interaction with the donor orbital. Ideal geometry is found in bridgehead amine derivatives such as **7.11** and **7.12**: these fragment to form cyclic imines, which are further hydrolysed to C=O groups under the conditions.

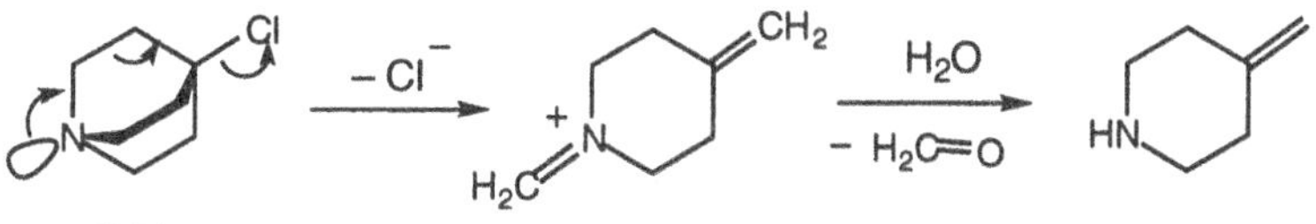

7.11

7.12

If the geometry does not allow good overlap the reaction simply does not happen. Thus the amine tosylate **7.13** does not fragment. The lone pair on nitrogen can occupy either axial or equatorial positions, but it cannot get antiperiplanar to the C—C bond antiperiplanar to C—OTs because this position is taken by a ring bond. (The relevant antiperiplanar bonds are in bold in **7.13**.) The result is a mixture of E1 and S_N1 reactions, as for the corresponding decalin **7.13a**, which reacts 2-3 times faster. Fragmentations like those shown for **7.11** and **7.12** are much faster than the E1 and S_N1 reactions of the carbocyclic analogues.

7.13 　　　　　　　　**7.13a**

7.2 Rearrangements

Whole books have been written about rearrangement reactions, including one in this series, so the discussion here is limited to general points about the stereoelectronic effects involved.

Although various sorts of rearrangements are known, the most common and important are 1,2-shifts, as depicted above (**7.5** → **7.6**). The migrating group can be alkyl, aryl or hydride (*not* a proton: H migrates with its pair of electrons), and, as before, the acceptor orbital can be σ*, π* or a vacant p-orbital.

The migration of an alkyl group involves retention of configuration at the migrating centre C_γ, as indicated in **7.6**, and as prescribed by the orbital-orbital interactions involved. The simple orbital picture which gives the best feel for the stereoelectronic situation is illustrated in Fig. 7.3, representing the interconversion of two carbocations **7.14** and its isomer **7.14'**. As usual the donor and acceptor orbitals should be coplanar, and specifically antiperiplanar to each other when the acceptor is σ*. The stereochemistry at the other two centres is what would be expected, ranging from complete inversion to complete racemisation if a carbocation is a full intermediate. The reaction is in principle reversible (and thus symmetrical in relevant cases, sometimes leading to unexpected racemisations).

The structure **7.15** (*not* an intermediate, possibly a transition state) is stabilised by a two-electron bonding interaction between the three centres involved. Each centre becomes electron-deficient in the course of the reaction, as indicated by the favourable effects of electron-donating substituents, but the migrating group retains its stereochemistry throughout.

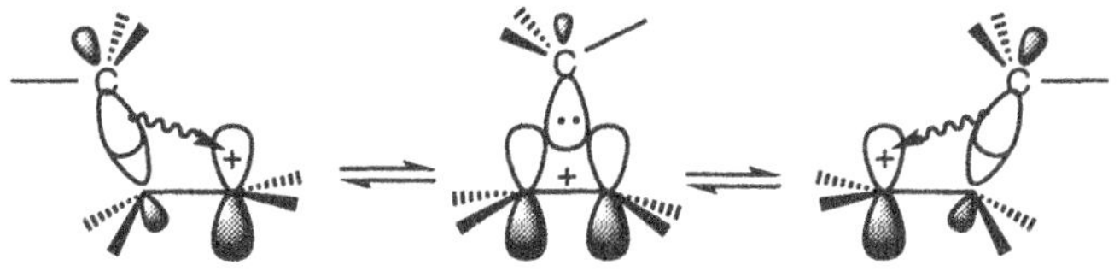

| 7.14 | 7.15 | 7.14' |

Fig. 7.3

The stereochemistry is simplest in migrations across a double bond, and the requirements are more restrictive than for eliminations. Here the migrating group must be antiperiplanar to the leaving group. Thus the Beckmann rearrangement gives an intermediate **7.16** with the positive charge delocalised onto the adjacent nitrogen atom:

Note that the orbitals directly involved in this rearrangement all lie in the plane of the π-system. This includes the lone pair on nitrogen, which evidently finds a way to interact with the developing carbocation at C_β, unpromising though the geometry looks.

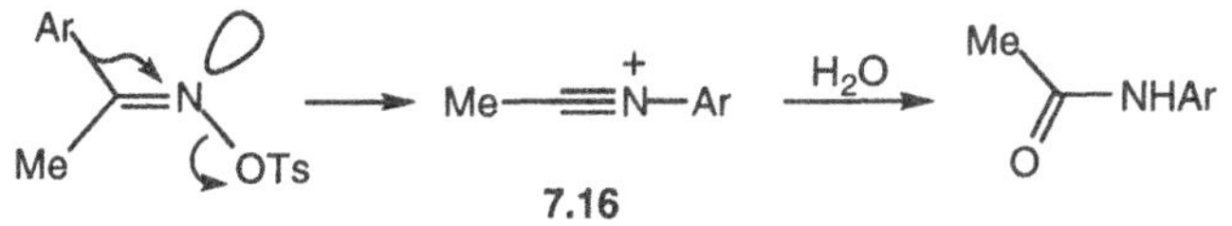

The same structural features which are particularly effective in stabilising a positive charge at C_γ are equally effective at C_β. So we are not surprised that two of the reactions of this sort which are familiar to every organic chemist involve stabilisation of C_β by an adjacent hetero-atom donor. Thus, although the contribution of the oxime nitrogen is essential to the Beckmann rearrangement, additional stabilisation of C_β by an oxyanion makes the reaction easier still. This is the basis of an important class of reactions in which amide derivatives (**7.17**) with a leaving group on nitrogen are converted to isocyanates (**7.18**), and thence to amines or their derivatives with the original acyl group R on nitrogen (see Problem 7.4).

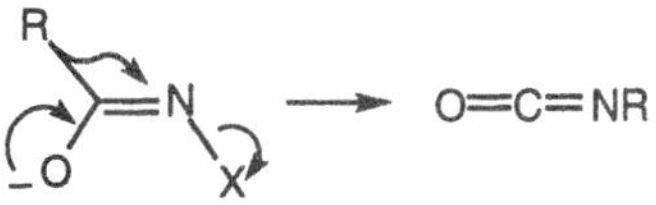

A comparable reaction in saturated structures is the pinacol rearrangement. There are no stereoelectronic restrictions on OH or oxyanion donors in these reactions.

7.18   **7.19**

7.3 Suggestions for further reading

L. M. Harwood, *Polar rearrangements*. Oxford Chemistry Primer no. 5, Oxford, 1992 gives an excellent introduction to rearrangement reactions. For a seriously systematic summary of rearrangement and fragmentation reactions in synthesis see J. B. Hendrickson, *J. Am. Chem. Soc.*, 1986, **108**, 6748.

7.4 Problems

7.1 Suggest why 2-acetylcyclohexanone (**E.1**) + NaOH behaves differently from 2-acetylcycloheptanone (**7.9**, page 79). Write down the mechanism and the major products for the fragmentation reaction of 2-acetylcyclohexanone.

7.2 Write a mechanism for the ethoxide-catalysed conversion of **E2** to **E3**, identifying the antiperiplanar relationships involved in the fragmentation step.

E.1   **E.2**   **E.3**

7.3 Explain the contrasting reactions of the isomers **E.4** and **E.5**, under identical conditions.

(a)   **E.4**   (b)   **E.5**

7.4 Look up (if necessary) the Schmidt, Curtius, Lossen and Hofmann (amide) rearrangements, and show how the mechanism shown as **7.17** (page 80) can be applied in each case.

8 Radical reactions

Stereoelectronic effects on radical reactions arise from geometrical preferences for optimal orbital-orbital interactions, in the same way as the effects on ionic reactions discussed in previous chapters of this book. The geometries of the relevant orbital-orbital interactions are the same as in the corresponding ionic reactions because the orbitals involved are the same. What is different is that the interacting orbitals contain an odd number of electrons.

Stereoelectronic effects on reactivity are generally simpler than for ionic reactions because the reactions are simpler. To illustrate the point consider the radical chlorination of an alkane.

$Cl-Cl$	$\overset{h\nu}{\rightarrow}$	$2\ Cl\bullet$		*Initiation*
$R-H\ +\ \bullet Cl$	$\rightarrow$	$R\bullet\ +\ HCl$		*Propagation*
$R\bullet\ +\ Cl-Cl$	$\rightarrow$	$RCl\ +\ \bullet Cl$		*Propagation*
$R\bullet\ +\ \bullet R\ (\bullet Cl)$	$\rightarrow$	$R-R$	$(RCl,\ Cl_2)$	*Recombination*

The products of the reaction arise in the propagation step or steps, which have to compete with the very fast recombination process. Recombination is generally diffusion-controlled - so fast that reaction occurs every time two radicals (with opposite spins) meet. Thus only very fast - which in practice means simple - propagation reactions are possible. Rearrangements are rare, and radical substitutions, like those in the propagation steps above, take place almost exclusively at singly bonded centres: i.e. at H or halogen (occasionally also at O or S). We may assume that the orbital-orbital interactions involved are subject to the usual stereoelectronic effects, but these reactions have no observable stereochemical consequences at the centres concerned.

The rare examples of radical substitution at carbon almost invariably involve the weak $C-C$ bonds of cyclopropanes. The addition of chlorine to 1,1-dichlorocyclopropane, for example (Fig. 8.1), is thought to involve an S_H2 process as the first propagation step, and has been shown to go with inversion of configuration: like an S_N2 reaction, and no doubt for basically the same reasons. (The second step is not stereospecific, as is to be expected if the radical **8.1** is a full intermediate.)

Radical substitution reactions are conveniently written with single-headed arrows (an equivalent set can be drawn in the reverse direction).

$$R\!\!-\!\!H\ \frown\ \bullet Cl$$

The shorthand for such a reaction is S_H2; similar to S_N2, with the H standing for Homolytic.
Radical reactions, which do not usually involve charged groups, are typically insensitive to solvent effects. Solvation is therefore not an important complicating factor in most radical mechanisms

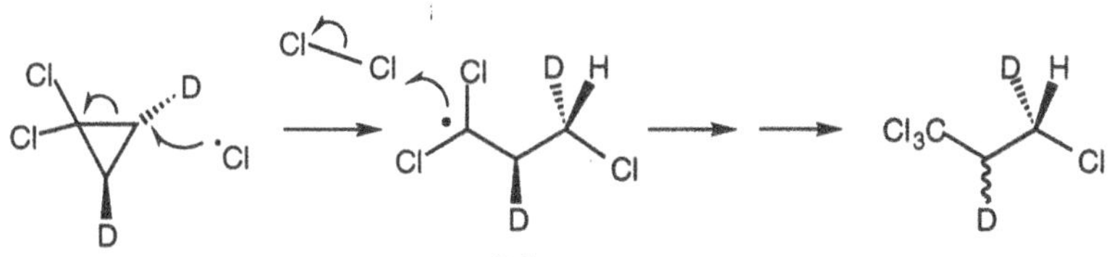

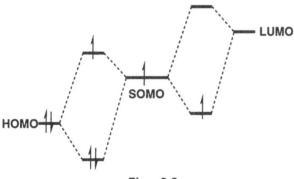

Fig. 8.1

This is an exceptional case, and it is no coincidence that the bond being broken is particularly weak (the formation of **8.1** involves the cleavage of a weak C—C bond and the formation of a stronger C—Cl bond). Reaction is further favoured in this case because the radical formed is stabilised by the two chlorine atoms.

8.1 Stability of radicals and reactivity

Because radicals react with low activation energies transition states for their reactions are close in structure to the radical intermediates (the Hammond Postulate: see page 42). So the major factor controlling reactivity in radical reactions is the stability of the reactant and product radicals concerned. Here there is an important difference from polar intermediates: *any* feature which stabilises either a carbocation or a carbanion will stabilise a radical. This is because a singly occupied molecular orbital (SOMO) can be involved in bonding interactions with either filled or vacant orbitals. The principle involved is illustrated in Fig. 8.2.

LUMO

SOMO

HOMO

Fig. 8.2

Fig. 8.2 shows two separate orbital interaction diagrams, showing how the singly-occupied molecular orbital (SOMO) of a radical can interact productively with either a high-energy occupied orbital (HOMO, right to left interaction) or with a sufficiently low-lying vacant orbital (LUMO, left to right). In practice the SOMO interacts with every filled or vacant orbital within range, but the interactions with the HOMO and the LUMO will be the strongest and most significant.

Consider first the interaction of the SOMO with the filled orbital, labelled HOMO. Two of the three electrons involved can be accommodated in the resultant bonding orbital, while the third must go into the new antibonding orbital. The net result is bonding, because two electrons are now in a lower energy orbital, while only one has gone up in energy. The picture is simpler when the SOMO interacts with a vacant orbital (LUMO in Fig. 8.2). Now there is only a single electron to be accommodated. It will go into the new bonding orbital, which is lower in energy than the original SOMO; so again the net interaction is bonding.

As a result radicals are stabilised, and thus formed faster, in systems where the unpaired electron is in an orbital (the SOMO) which can interact with either a donor or an acceptor orbital, (or with both at the same time). This can be p, π or σ in type. Some reactions in which stabilised radicals are formed are shown in Fig. 8.3. The geometries of the interactions are not new, though at first sight some of the results are surprising. The H atom removed from an alcohol, for example, comes not from oxygen but from the carbon atom next to it (**8.2**). The radical formed is the one with the most efficient delocalisation, with the SOMO interacting with a lone pair on oxygen (**8.3**). The oxygen of a carbonyl group is also effective enough as a lone-pair donor to stabilise an acyl radical (**8.4**). Other examples in Fig. 8.3 are more familiar; but note that H is removed equally easily from CH_2 next to C=C and C=O.

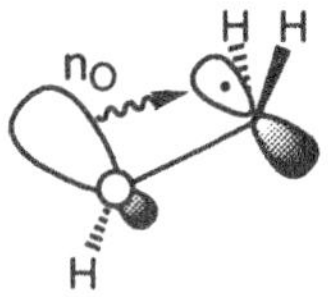

8.3

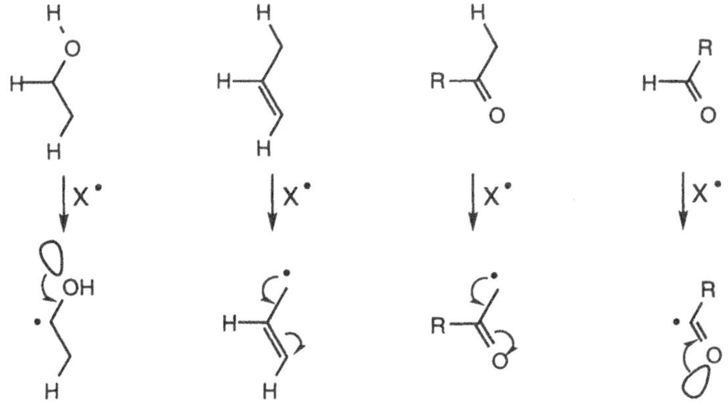

8.2

8.4

Fig. 8.3

In the absence of strong delocalisation the structure of an alkyl radical can conveniently be thought of as intermediate between that of a carbocation and a carbanion. There is relatively little difference in energy between planar and pyramidal forms, so inversion at a radical centre is rapid and stereochemistry is not preserved (Fig. 8.4). Similarly, radicals are formed at bridgeheads, where they must be pyramidal, more easily than are carbocations (though more slowly than in acyclic systems).

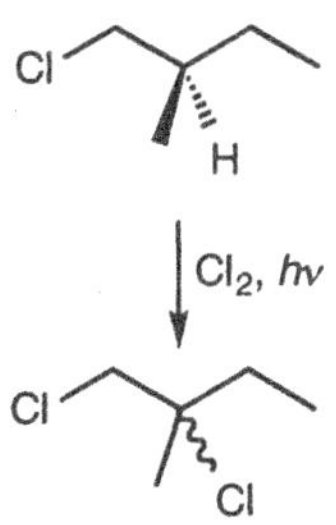

Fig. 8.5 compares relative reactivities towards the t-butoxy radical of four different secondary C—H bonds and an efficient hydrogen atom donor much used in synthesis. (The weak tin-hydrogen bond means that Bu_3SnH reduces most radicals faster than they react with other organic compounds.)

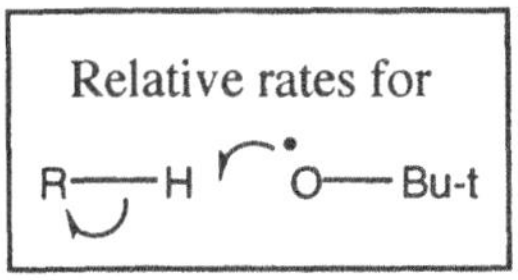

Fig. 8.4

Relative rates for
R—H ᷍O—Bu-t

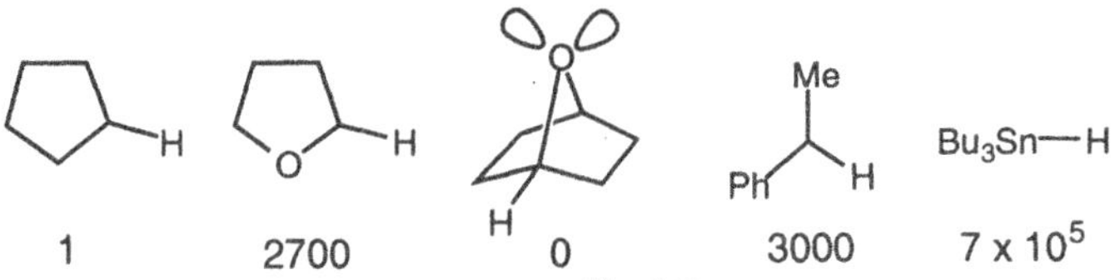

1   2700   0   3000 7×10^5

Fig. 8.5

Note that the effect of an adjacent oxygen is comparable with that of a
benzene ring; but that it requires efficient overlap of the oxygen lone pairs
with the radical (**8.3**), in the same way as stabilisation of a cationic centre at
a comparable position (see above, page 41). Thus the bridgehead C—H bond
(Fig. 8.5) does not react at a significant rate; similarly axial C—H bonds of
1,3-dioxans (**8.5a**), with antiperiplanar lone pairs, react some ten times
faster than those of the equatorial isomers (**8.5e**).

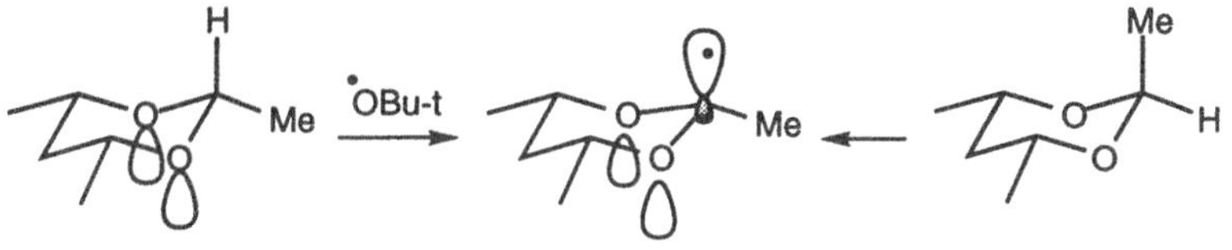

8.5a   **8.5e**

8.2 Intramolecular reactions

The easy abstraction of hydrogen from a C—H bond is unique to radical
chemistry, and the intramolecular version is one of the commonest of all
radical reactions. There is a strong preference for reaction through a six-
membered cyclic transition state, even when other ring sizes would give more
stable radicals. Clearly there is no stereoelectronic barrier to the reaction,
though we saw in Chapter 5 (page 36) that the linear transition state for an
intramolecular 'S$_N$2' reaction cannot be accommodated in a 6-ring. The
difference is that even though a linear transition state (Fig. 8.6) may be
preferred for radical substitution at H (to minimise steric and electrostatic
interactions), the preference is not a strong one. The C—H σ-bonding orbital
(Fig. 8.6) is not as strictly directional along the axis at the H-end as a C—C
bond, because of the symmetry of the original hydrogen (1s-orbital)
component.

Fig. 8.6

The 1,5-hydrogen transfer is therefore rapid, and can take advantage of the
usual advantage of an intramolecular reaction, because the non-linear
transition state (Fig. 8.7) can be accommodated without significant strain in
a 6-membered ring. (1,4-hydrogen transfers, which must go through
5-membered transition states, are much slower: presumably the transition
state can tolerate only so much bending.)

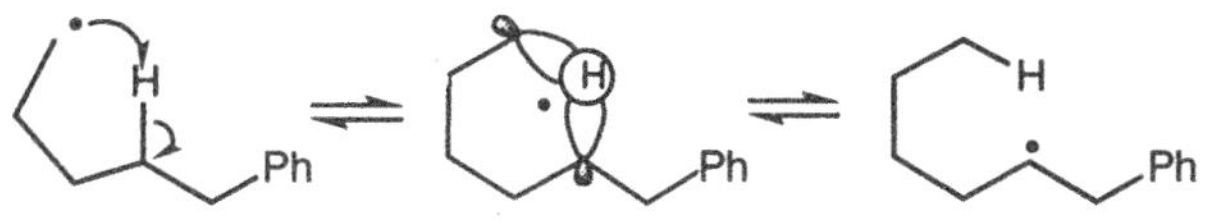

Fig. 8.7

8.3 Reactions involving diradicals

The reactions of diradicals, with two radical centres in the same molecule, are complicated by the problem of spin multiplicity - whether the two have paired or unpaired spins. If we ignore this complication the chemistry is very simple. The two radicals do their best to interact with each other, and the usual result is cyclisation by recombination.

1,4-Diradicals are close enough to interact through bonds, and will fragment if the geometry is right. Since the orbitals involved are the same, albeit half occupied in two cases, the stereoelectronic requirements are the same as for ionic fragmentations (see above, page 78). The antiperiplanar geometry is preferred (Fig. 8.8)

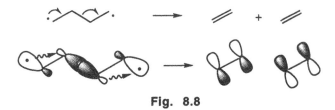

Fig. 8.8

1,3-Diradicals prefer to cyclise rather than to rearrange. They occur most commonly as intermediates in the addition of 1,1-diradicals - triplet carbenes - to alkenes, and cyclisation is typically fast enough to be in competition with rotation about the new C—C single bond (**8.6**), as shown by the loss of the alkene stereochemistry.

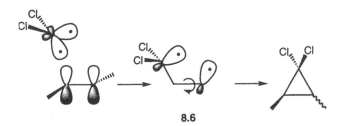

8.6

One of the easiest ways of generating 1,2-diradicals is by the photochemical activation of π-systems. When a photon of the right wavelength is absorbed an electron is promoted, usually from the HOMO, to a higher electronic energy level, most often the LUMO. The result is to break the bond concerned, since this puts an electron in the antibonding orbital, cancelling the effect of the remaining bonding electron. If a σ-bond were involved two separate radicals would be produced, but in the case of a π-bond these are held together by the σ-framework. The preferred geometry of the excited state of a simple alkene has the two half-filled orbitals perpendicular to each other (**8.7**), so the relaxation process can produce either cis or trans alkene. This is often the most convenient route to the other geometrical isomer.

8.7

The perpendicular orbitals in **8.7** are drawn for clarity as p-orbitals. They are of course π and π*.

The photochemistry of carbonyl compounds also involves diradicals, often of more than one sort. It is described in the following section.

8.4 Photochemistry of C=O compounds

The photochemical activation of a saturated ketone or aldehyde is of special stereoelectronic interest. The LUMO is $\pi^*_{C=O}$, as expected, but the HOMO is not $\pi_{C=O}$ but one of the lone pairs on the carbonyl oxygen. This has two interesting consequences. First, the resulting excitation (n_O-$\pi^*_{C=O}$) is formally forbidden, since the orbitals concerned are orthogonal, and the lone pairs lie in the node of the π-system. This has the effect in practice of making the absorption (near 330 nm) particularly inefficient. Nevertheless, excitation does happen.

The second effect follows from the first: once excitation has occurred the two radical centres behave - and in particular react - independently, because the two singly occupied orbitals are orthogonal (**8.8**).

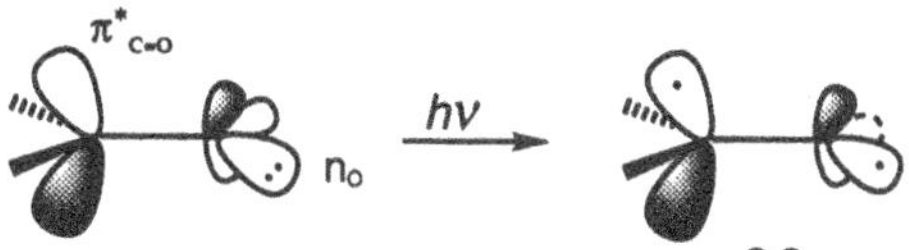

8.8

Thus radical additions can be initiated either by the more electrophilic oxygen radical centre, interacting with the HOMO of an alkene (and going through the more stable 1,4-diradical **8.9**); or by the carbon-centred radical, interacting with an alkene LUMO. In each case the product is a 4-membered ring, formed by way of the more stable 1,4-diradical (**8.10** in the second case); but the direction of addition is controlled by the HOMO/LUMO interactions.

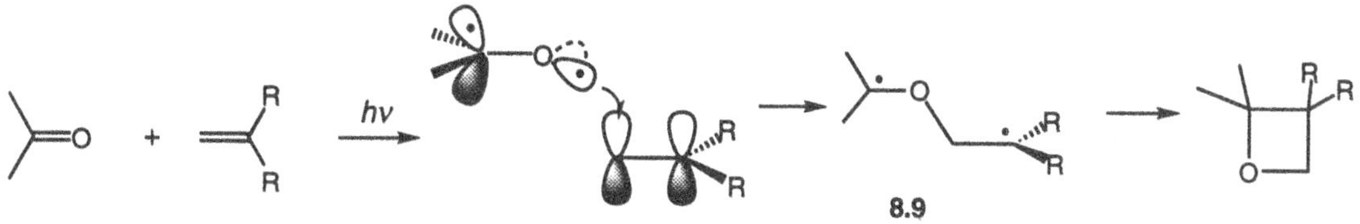

8.9

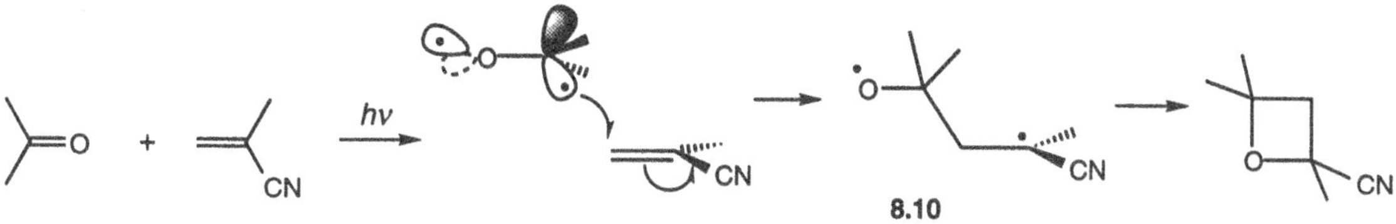

8.10

Note that both processes described in this section, initiated by the sp²-hybridised radical centred on oxygen, happen also to the radical cations produced in electron-impact mass spectrometry.

Intramolecular reactions

In the absence of an easy intermolecular reaction, photochemically excited C=O compounds undergo rapid intramolecular processes, which end up with the two radical centres terminating each other. Two of these processes are of major importance, and both are initiated by the reactive oxygen radical centre.

The planar relationship of the unpaired electron on oxygen and the bonds at the carbon centre allows the eliminative cleavage of the (presumably) antiperiplanar C—C bond (8.11). Further reactions include a second C—C cleavage to produce CO, followed by recombination of the two alkyl radicals.

8.11

In competition with this α-cleavage is the rapid abstraction of a hydrogen atom, from any C—H bond available at the correct distance (remember that a 6-membered cyclic transition state is required). This produces a 1,4-diradical (*cf.* 8.9), which can cyclise to form a cyclobutanol (8.12) or fragment if the overlap is favourable in an accessible conformation. This reaction produces two π-systems, one of them an enol (8.13). This rapidly ketonises, but can be observed directly under the right conditions.

8.12

8.13

8.5 Suggestions for further reading

A useful introduction to radical reactions in synthesis is Chapter 2 of C. J. Moody and G. H. Whitham, *Reactive Intermediates* Oxford Chemistry Primer no. 8, Oxford, 1992. See also B. Giese, *Radicals in Organic Synthesis: Formation of Carbon-Carbon Bonds*, Pergamon, Oxford, 1986. For a more mechanistic treatment see Chapter 9 of T. H. Lowry and K. S. Richardson, *Mechanism and Theory in Organic Chemistry*, Harper and Row, New York, 1987.

Answers to problems

3.2 **3.15a"** would suffer serious steric hindrance.

3.3 Only **A.3** has the preferred gauche-gauche conformation (**3.16**).

3.4 **3.33** has both disfavoured *trans, trans* acetal conformations and
unfavourable gauche interactions across the central ring-bond.

3.5 Steric effects favour **A5e**, the gauche effect favours **A5a**, and is greater
for the more electronegative OSO_2Me group.

4.1 The lone pair cannot conjugate, mainly because N is protonated in acid!

4.2 Stabilisation of the carbanion $(F_3C)_3C^-$ by σ-delocalisation into up to
nine C–F bonds (*cf.* **4.11**, p. 26) is not available to F_3C^-.

5.2 Dealkylation is expected to be intermolecular (NOT 7-endo-tet).

5.3 Displacement of chloride must occur in the plane of the π-system, so the
transition state cannot benefit from benzylic stabilisation.

5.4 5-exo-tet is favoured, to give **C8** and **C9**.

5.5 E^+ adds to C=C to generate C^+ β to Si in each case, thus taking
advantage of the σ-donor properties of the Si–C bond.

5.6 **C6** prefers 5-exo-tet, but making one centre allylic shifts the balance to
6-exo.

5.7 The 3 neopentyl groups on Sn give maximum steric protection to the
CH_2–Sn bonds, and direct attack to the desired bond to the chiral centre.

6.1 The carbanion **6.13** lives longer under more basic conditions, and
rotation about the former double bond can occur. Protonation retains its
stereoelectronic preference.

6.2(a). The CF_3 group must stabilise adjacent C^- by (σ-delocalisation) more
effectively than does the benzoyl group acting as a π-acceptor.

6.2(b). Consecutive elimination-addition reactions, via alkyne intermediates,
must be involved.

6.3 As described in the text (pp. 57-58) for allylsilanes: the Me_3Si group
directs E^+ to the opposite face of C=C, for both steric and
stereoelectronic reasons.

6.4 An equatorial H is antiperiplanar to a $(C–N^+)$ ring bond in the chair
conformation of the 6-membered ring, allowing a favourable E2 process
not available to the 5-membered ring compound.

7.1 Addition to the C=O group of the 6-membered ring is favoured, by the
all-staggered conformation of the chair form of the addition intermediate.

7.2 Ethoxide adds to **E2** to give **E6**, which has a series of antiperiplanar
σ-bonding orbitals linking the OTs leaving group with the lone pair
on O^-.

3.15a"

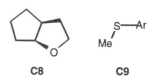

C8   **C9**

7.3 **E.4** can fragment as shown (the $C=N^+$ double bond produced is reduced
by the borohydride). **E.5** has only C–H antiperiplanar to the bond to the
axial OTs leaving group.

Index

Printed and bound by CPI Group (UK) Ltd, Croydon, CR0 4YY